Economics of crops in developing countries No 1

The economics of coffee

The series 'Economics of crops in developing countries' is the result of a research project undertaken by the Department of Development Economics of the Agricultural University, Wageningen. The monographs in this series are intended to fill the gap between agronomic documentation and aggregate data on production and trade. The following titles are in preparation:

2. The economics of oil palm: H.A.J. Moll
3. The economics of maize: J.A. Kool
4. The economics of cotton.

Economics of crops in developing countries No 1

The economics of coffee

J. de Graaff

Pudoc Wageningen 1986

Ir J. de Graaff is an agricultural economist, who graduated from the Agricultural University, Wageningen, in 1974. He worked for the Food and Agriculture Organization of the United Nations in Tunisia, Kenya and Jamaica and at the headquarters in Rome. He was employed by the Department of Development Economics of the Agricultural University from 1982 to 1984. In 1984, he joined the staff of the Royal Tropical Institute in Amsterdam. Since 1979, he has carried out several consultancy assignments in Africa and Asia.

CIP

ISBN 90 220 0900 9

Contents

Foreword

In 1982, the Department of Development Economics of the Agricultural University in Wageningen embarked upon a project aimed at the production of a series of monographs on the economics of tropical and subtropical crops. Each monograph was to cover the aspects of production, processing and trade of a specific crop, with the intention to fill the gap between the existing ample agronomic documentation, and the generally broad and aggregated data on production and trade. The studies are based on information from professional publications and statistical sources, complemented by more detailed field studies in major producing countries.

The first phase of the project has covered four crops with different ecological and economic characteristics, i.e. coffee, cotton, maize and oil palm. Each study will devote attention to the following aspects:

- ecology in various regions
- farm types, farming systems and production
- the position of the crop in relation to other crops with either similar uses or similar ecological requirements
- the stages of processing and marketing between farm gate and consumer or importer
- supporting services for production, processing and marketing
- national institutions, programmes and policies
- international trade.

We hope that this comprehensive approach will make the monographs useful as reference manuals for institutions and people involved in policy-making and field work.

Professor Dr F.P. Jansen
Department of Development Economics,
Agricultural University, Wageningen

Preface

Coffee is one of the most important agricultural export products. It is grown almost exclusively in developing countries. Many of these countries depend heavily on coffee for their foreign exchange.

This book describes and analyses the coffee sector and its role in the national economy of producing countries. Emphasis is given to micro-economic aspects of production, processing and marketing, and macro-economic aspects of the sector as a whole. Moreover, attention is paid to physical and institutional factors that have a profound influence on economic performance, such as ecology, organization of services and policy matters.

Eight countries in Latin America, Africa and Asia have been selected for analysis. Together, these countries are representative of the situation in about fifty producing countries. For each of the selected countries, the coffee sector is reviewed and analysed in sections dealing with the general background; the ecological conditions; the availability of production factors; the production, processing and marketing with methods applied and costs; the provision of supporting services and government policies. The findings in these eight countries provide the basis for conclusions about the potential and limitations of the coffee sector in producing countries.

The description of the actual situation in producing countries and the subsequent comparative analysis offer general and specific insights into the coffee sector to readers with different fields of interest. Specialists will find it useful to have a comprehensive picture of the sector as a whole and to compare specific situations in several producing countries; planners may use data and parameters on the various production stages for planning purposes; policy makers may compare the performance of one or more producing countries with the actual situation in their country and identify potential areas for improvement.

The book consists of two parts. Chapter 1 of Part I provides a general review of present knowledge about coffee and its derived products, and serves as an introduction for those who are not acquainted with production and trade in coffee. The comparative analysis based on the eight country studies is presented in Chapter 2. Individual country studies constitute Part II, Chapters 3-10. These studies follow a standard outline to facilitate comparison of specific aspects.

The data in this book refer to the situation in 1982, and prices and parameters will change with time. However the framework of analysis employed, together with the relatively constant basic factors and technical coefficients offer the individual user the possibility of updating sections according to his needs.

Acknowledgments

A study like this requires a vast amount of information about the many aspects of production and trade. Therefore I am grateful to a large number of specialists, who were prepared to spend their valuable time discussing the various aspects of coffee, and to the many institutions that made available reports, studies, statistical and financial data, which are generally not available in libraries. Without their participation, this study could not have been accomplished.

In the first stage of the research, valuable assistance was received in establishing contacts and in collecting information from staff members of various institutes in the Netherlands: Ministeries of Agriculture and Fisheries, Economic Affairs and Foreign Affairs; several semigovernmental institutes, consultancy firms and trading companies; and from several national and international organizations in other European countries. In particular I would like to thank Messrs M. van de Steene, H.C.G. Hawkins and C.P.R. Dubois of ICO, London; Mr J.M. Gastellu of ORSTOM, Paris; Dr K.H. Friedrich of Institut für Grünlandwirtschaft, Brunswick; and staff members of the Agricultural Services and the Commodity & Trade Divisions and the Investment Centre of FAO, Rome.

For the visits to the producing countries, support in many aspects was received from the Netherlands Embassies and Consulates. I received much help from numerous institutes and many persons in those countries, but I would particularly like to thank the persons and institutes mentioned below for their kind assistance: Dr C.E. Fernandez and Mr P. Bornemisza of IICA and Mr R. McColaugh of ROCAP, and CATIE in Costa Rica; Dr G. Valenzuela Samper, Dr R. Lopez Alzate and other staff members of Federacafé in Colombia; Mr G.J.M. Terberg, Agricultural Attaché of the Netherlands Embassy, and Mr K. N'Guetta of SATMACI, Mr F. Ruf of GERDAT, the Caisse de Stabilisation and ORSTOM in Ivory Coast; Messrs P.B. Tchatat and M. Moreau of UCCAO and several staff members of DGRST in Cameroon; Mr E. Ruhigira of OCIR-Café and Mr J.H. Doyen of the World Bank in Rwanda; Mr F.N. Kiragu of the Coffee Board of Kenya, Mr H. Schmidt of the Ministry of Agriculture, Dr G.O. Mbaja of the Ministry of Cooperative Development and the Coffee Research Foundation in Kenya. For the country study of Brazil, contacts were established through Dr B. Eskes, EMBRAPA, and I would like to thank the Instituto de Economia Agricola, São Paulo, for their extensive information. I owe many thanks to Mr P.A.L de Rijk, who provided me with much material for the country study of Indonesia, and who commented on a first draft. Dr H.A.M. van der Vossen was kind enough to read and comment on the introductory chapter.

I am also much indebted to my former colleagues of the Department of Development Economics for their stimulating discussions, comments and general interest in the study. In particular, I wish to thank Mr H.A.J. Moll for pleasant collaboration on this project. The typing of successive drafts of this book was efficiently undertaken by Mrs C.O. van Overbeek-van Amersfoort and Mrs G.J. van Lieshout-Boshuizen.

Lastly I wish to thank Messrs R.J.P. Aalpol and J.C. Rigg of Pudoc for their advice and guidance about all editorial aspects.

Abbreviations

Almacafé	– Almacenes Generales de Deposito de Café (Colombia)
BAPPENAS	– National Panning Bureau of Indonesia
BETPA	– Bureau d'Etudes Techniques des Projets Agricoles (Ivory Coast)
CATIE	– Centro Agronomico Tropical de Investigación y Ensenanza (Costa Rica)
CBK	– Coffee Board of Kenya
CEDAR	– Centre d'Etude et de Développement de l'Arabusta
Cénicafé	– Centro Nacional de Investigaciones de Café (Colombia)
CFAF	– Communauté Financière Africaine Franc currency monetary unions of West-Africa (Ivory Coast) and Central-Africa (Cameroon)
Cicafé	– Centro de Investigaciones en Café (Costa Rica)
CIDT	– Compagnie Ivoirienne pour le Développement des Textiles
CIRES	– Centre Ivoirien de Recherches Economiques et Sociales
colones	– Costan Rican currency
CONICIT	– Consejo Nacional de Investigaciones Cientîficas y Tecnológicas (Costa Rica)
CRF	– Coffee Research Foundation
Cr$	– (Brazilian) Cruzeiro (currency)
CSSPPA	– Caisse de Stabilisation et de Soutien des Prix des Produits Agricoles (Ivory Coast)
DGRST	– Délégation Générale à la Recherche Scientifique et Technique (Cameroon)
ECLA=CEPAL	– Economic Commission for Latin America
EMBRAPA	– Empresa Brasileira de Pesquisa Agropecuaria
FAO	– Food and Agriculture Organization of the U.N.
Federacafé	– Federación Nacional de Cafeteros (Colombia)
f.o.b.	– free on board
Fonader	– Fond National de Developpement Rural (Cameroon)
f.o.r.	– free on rail
FRw	– Franc Rwandais (currency)
GDP	– Gross Domestic Product
GERCA	– Grupo Executivo de Rationalizaçao de Caféicultura (Brazil)
GNP	– Gross National Product

GVC	– Groupement à Vocation Coopérative (Ivory Coast)
IBC	– Instituto Brasileiro do Café
ICA	– International Coffee Agreement
ICO	– International Coffee Organization
IICA	– Instituto Interamericano de Cooperación para la Agricultura
IMF	– International Monetary Fund
IRA	– Institut de la Recherche Agronomique (Cameroon)
IRCC	– Institut de Recherches du Café et Cacao et autres plantes stimulantes (Ivory Coast)
ISAR	– Institut des Sciences Agronomiques du Rwanda
KIT	– Koninklijk Instituut voor de Tropen (Royal Tropical Institute)
KPCU	– Kenya Planters Cooperative Union
KShs	– Kenyan Shillings (currency)
MAG	– Ministerio de Agricultura y Granaderia (Costa Rica)
ORSTOM	– Office de la Recherche Scientifique et Technique Outre Mer (France)
NWCA	– North West Cooperative Association, Ltd (Cameroon)
OCIR	– Office des Cultures Industrielles de Rwanda
Oficafé	– Oficina del Café (Costa Rica)
ONCPB	– Office Nationale de Commercialisation des Produits de Base (Cameroon)
pesos	– Colombian currency
PIK	– Coffee Intensification Project (Indonesia)
Prodesarrollo	– Programa de Desarrollo y Diversificación (Colombia)
Promecafé	– Programa Cooperativa para la Producción y Modernización de la Caficultura en Méjico, Centro America y Panama
ROCAP	– Regional Office for Central America and Panama (USAID)
Rp	– Rupiah (Indonesian currency)
SATMACI	– Société d'Assistance Technique pour la Modernisation Agricole de la Côte d'Ivoire
SCIP	– Smallholder Coffee Improvement Project (Kenya)
SOCOPAO	– International trading firm (Cameroon)
STICA	– Servicio Técnico Interamericano de Cooperación Agrícola
UCR	– University of Costa Rica
USDA	– United States Department of Agriculture
Zapi (de l'Est)	– Société Régionale du Développement des Zones d'Action Prioritaires Integrées de l'Est (Cameroon)

Part I. Basic facts and economic analysis

1 Introduction to coffee

1.1 General aspects

1.1.1 Botany

1.1.1.1 Classification

Coffee is the major genus of the family Rubiaceae, which includes some 400 genera and 500 species, mostly trees and shrubs, mainly found in the lower regions of the tropical rain forest. The family includes some ornamentals, tannins and some drugs of which *Cinchona* spp. are the major source of quinine (Purseglove, 1977).

The genus *Coffea* ranges from slender sprawling plants to robust trees with clean trunks and spreading heads growing 10-20 m high. It was classified botanically by Linneaus in 1737, and the best available system for nomenclature of the species is that of Chevalier (1929-1942). Of the four main sections of the genus, as classified by Chevalier: Eucoffea, Argocoffea, Mascarocoffea and Paracoffea, the first three are exclusively native to Africa, whereas most of the representatives of the fourth section are indigenous to tropical South-East Asia (e.g. *C. bengalensis*). Mascarocoffea, mainly found as wild trees in Madagascar, contains hardly any caffein.

The section Eucoffea includes the economically most important species:

- *C. arabica*, in the trade referred to as arabica and accounting for 76 % of the world's commercial coffee and about 9 thousand million adult cultivated plants
- *C. canephora* Pierre ex Froehner (syn. *C. robusta* Linden), in the trade referred to as robusta, with 23 % of the world coffee and about 3.5 thousand million adult cultivated plants
- *C. liberica* Hiern. (W. Africa), with 1 % of world coffee.

The best known varieties of *C. arabica* are 'Typica' and 'Bourbon' (from Réunion). A whole series of important strains and cultivars have been developed from the above, such as Caturra in Brazil and Colombia, mundo novo in Brazil, the hybrid *C. tico* in Central America, the 'Blue Mountain' cultivar from Jamaica, and several series of varieties, developed from 'Bourbon' in East Africa and India. One of the major criteria for selection was resistance to leaf-rust or coffee berry disease.

The best known varieties of *C. canephora* are 'Kouilou', widely cultivated in

Ivory Coast (Petit Indénié); the smaller plants 'Nana' and 'Nganda'; and the most commonly known variety 'Robusta', which has been intensively selected in Indonesia and the Congo (Haarer, 1962).

Like *C. canephora*, *C. liberica* is of West African origin, and exhibits a wide variability because of its self-sterility.

1.1.1.2 Morphology

Coffea arabica originates from Ethiopia, and often grows into a multistemmed shrub-like tree, averaging 4 to 5 m in height, but may, in the wild state, reach 10 m. It has a kind of taproot that forms cluster-like ramifications. A well known botanical feature of coffee, important for pruning, is its two types of stem growth (dimorphic branching). Part grows vertically and part horizontally. The orthotropic vegetative parts consist of the trunk and cental axis, some upright shoots from axillary buds that normally remain dormant, and the (more or less) horizontally growing branches on which fruit is produced. The leaves are oval, dark green on the upper surface and a lighter colour underneath. They are about 15 cm × 5 cm and 22 cm × 8 cm for arabica and robusta coffee, respectively. In the axils of the leaves on the one-year and older branches (not the main stem), small white flowers arise in clusters.

C. arabica is a tetraploid (2 n=44) and is self-fertile. Some natural cross-pollination occurs, effected by insects and wind. The ovary develops into a globular or oval drupe, normally containing two seeds. It has a length of 14-18 mm and a diameter of 10-15 mm. It is usually called a cherry or a berry, although botanically not correctly so. The fruits take 7 to 9 months to mature. When mature, the skin is red (for some varieties yellow), covering a slippery sweet and mucilaginous pulp. Inside the fruit, the two seeds (coffee beans) lie with their flat sides together. Each of the two coffee beans is covered by a loose, thin and yellowish skin (parchment), with a coating of thin slimy mucilage. Underneath that skin is a thin and closely fitting membraneous tegument, known as the silver skin (Fig. 1.1). These layers have to be removed before roasting, by a hulling operation. The beans of *C. arabica* are 9-12 mm long, 6-7 mm wide and 3-4 mm thick, and weigh about 0.15-0.20 g. The average weight ratio of cherries to clean coffee beans is 5.5 : 1; and clean coffee contains about 2200 beans per kilogram.

The other, diploid, coffee species, *C. canephora* and *C. liberica*, are self-sterile

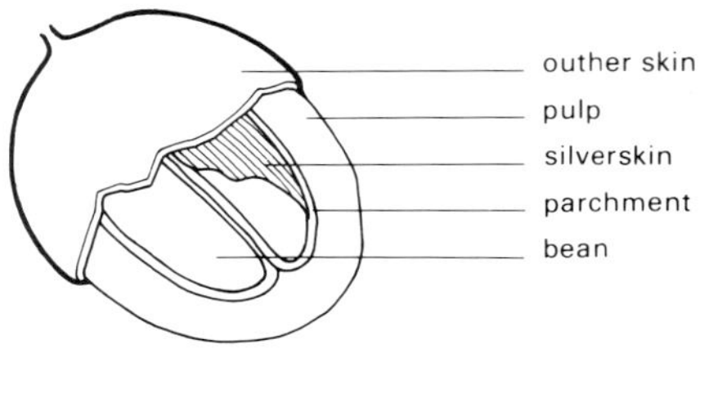

Fig. 1.1. Cross-section of coffee cherry.

and are mainly wind-pollinated. The fruits take longer to mature than do those of *C. arabica* (9-11 months for *C. canephora*). *C. canephora* is a robust glabrous shrub or small tree growing to 10 m in height. The tree is shallow-rooted, with the bulk of the feeding roots in the top 15 cm of soil. The trunk develops best in deep shade. The fruit is a rounded drupe of 12 mm diameter and colours crimson at ripeness, becoming black while drying on the tree until harvested. There is comparatively little pulp. *C. canephora* bears more fruits than *C. arabica*, 3-6 fruits developing from each inflorescence. The seeds are 8-9 mm long. The proportion of cherry to dried beans is approximately 4.5 : 1, and 1 kg clean coffee contains about 3300 beans.

Coffea liberica generally grows bigger than the other species, reaching between 5-17 m in height.

1.1.1.3 Propagation

Most commercial arabica coffee is grown from seeds, raised in nurseries. Provided seed is obtained from progeny-tested mother trees, vegetative propagation has little advantage, except for increasing clones and for topworking inferior plantations.

Robusta is more often multiplied vegetatively, for instance by cuttings from selected clones. The following selection criteria are used for planting material:
- suitability of the plant for local conditions of soils and climate
- strong and healthy growth
- regularity and volume of yield over a long period at a well defined density of plants
- good returns of marketable coffee
- maintenance of a certain minimum weight of beans
- resistance to diseases and pests
- caffein content.

1.1.1.4 Chemical composition

The coffee bean is a nut-like seed, largely of endosperm material and rich in stored foods. It is well filled with proteins, fats, sugars and mineral substances. Many studies have been made on the chemical contents of the coffee beans in different stages. The approximate composition of green beans and roasted beans is given in Table 1.1.

During the process of roasting, water evaporates, the sugars are caramelized and the cellulose is carbonized. The caffein content varies between species, and is approximately 1-1.5 % for *C. arabica*; 2.0-2.5 % for *C. canephora*; and 1.4-1.6 % for *C. liberica* and almost zero for mascaro coffee. The caffein content is not responsible for the bitterness: *C. arabica* with a caffein of 1.5 % may not show any and *C. liberica*, with the same proportion, may be markedly bitter. Of the nutri-

Table 1.1. The approximate composition (%) of green beans and roasted beans.

	Green beans	Roasted beans
Water	10-13	1-4
Protein	11-15	12-15
Fat	12-14	13-15
Sugar	8-10	0-1
Caffetanic acid	8-10	4-5
Caffein	1-2	1-2
Ash	4	4-5
Cellulose and allied subs.	32-46	–
Carbonized cellulose, etc.	–	53-65

Source: several publications

ents relevant for the application of fertilizers, coffee beans contain about 2.3 % N, 2.0 % K_2O and 0.4 % P_2O_5. At ripening stage, these nutrients are concentrated in the berries. Potassium and nitrogen are required by the crop at all times, potassium being the dominant fertilizer. Phosphoric acid is required chiefly at the beginning of the season.

1.1.2 Origin and distribution

Although coffee is mainly indigenous to the African continent and the surrounding islands, the use of coffee as a beverage was discovered in Arabia (15th Century) and the first major production countries were in Asia (Sri Lanka, Indonesia and India) in the 17th Century. Later coffee came into use in Latin America, where Brazil has predominated world coffee production since 1840.

The home of arabica coffee is the Ethiopian Massif, where it occurs naturally in forests between 1500 and 2000 m. The Arabs introduced it in some countries in the Middle East and in Sri Lanka (formerly Ceylon) before the Dutch brought it, probably from Yemen, to Java (Indonesia) in 1690. The story goes that the Burgomaster of Amsterdam sent vigorous progeny from one tree (*C. arabica* var. *typica*) originally from Java, to Louis XIV, who asked in his will that seeds from that 'tree' be distributed to all his tropical empire. This indeed happened. Through French Guyana, it reached Brazil in 1729; through Haiti and Martinique, it reached other Latin American and Caribbean countries (Wellman, 1961). The French also introduced coffee into African territories, and the other major arabica variety is called after the island Bourbon (now Réunion). From there, missionaries took it to various places including East Africa (1893; French Mission).

Much of the arabica coffee in Asia (Sri Lanka, Indonesia) was decimated by leaf-rust (*Hemileia vastatrix*) epidemics during the last two decades of the 19th Century. In Sri Lanka, coffee was then replaced by tea; in Indonesia, other coffee species were tried. Firstly *C. liberica*, which was also susceptible to leaf-rust, and then *C. canephora*, which proved to be resistant. *C. canephora* originated from Central Africa, chiefly from the belt between 10°N and 10°S of the equator, where it grows wild in the equatorial forests. Africans had planted it on a small scale before the arrival of the Europeans, and they collected beans from wild trees. The species name *C. canephora* was given by Laurent in 1895, but in the trade it became known as robusta coffee. It is now widely distributed in tropical Africa and Asia (not in Latin America), where it is grown successfully at lower altitudes less

Table 1.2. Production of green coffee for important production countries over the last three decades.

	Period					
	1949/51	1959/61	1969/71		1979/81	
	(1000 t)	(1000 t)	(1000 t)	(%)	(1000 t)	(%)
World	2191	4264	4262	100	5168	100
S. America	1478	2767	1885	44	2425	47
Brazil	1073	2171	1197	28	1403	27
Colombia	336	463	483	11	748	14
Ecuador	21	42	60	1	82	2
Peru	6	33	68	2	98	2
Venezuela	39	54	60	1	61	1
NC. America	365	569	733	18	918	18
Costa Rica	22	55	82	2	109	2
El Salvador	68	105	139	2	158	3
Guatemala	58	101	125	3	162	3
Haiti	36	36	31	1	32	1
Honduras	14	22	39	1	80	2
Mexico	64	116	182	4	216	4
Africa	271	741	1278	30	1198	23
Angola	46	142	216	5	44	1
Cameroon	9	34	90	2	102	2
Ethiopia	23	91	172	4	191	4
Ivory Coast	51	138	243	6	292	6
Kenya	11	28	57	1	84	2
Madagascar	33	48	63	1	84	2
Uganda	33	106	215	5	124	2
Zaire	21	56	71	2	81	2
Asia	76	184	340	8	576	11
India	21	55	82	2	130	3
Indonesia	40	93	173	4	253	5
Philippines	4	23	48	1	33	1
Oceania	1	3	25	1	49	1
Papua New Guinea	1	3	25	1	49	1

Source: FAO (1952-1983)

suitable for *C. arabica*, and in areas where *Hemileia* poses a serious problem.

The world trade in coffee developed as follows:

1720	90 t	(mainly from Java)
1770	320 t	(mainly from Asia, first amounts from Brazil)
1820	90 000 t	(half of which from Brazil)
1870	450 000 t	(not much any more from Asia)
1920	1 600 000 t	(two-thirds from Brazil)
1970	3 200 000 t	(half from Brazil & Colombia; 25 % from Africa).

The recent development of coffee production in the major production countries is shown in Table 1.2.

1.1.3 Uses and competitive product

It is almost certain that coffee fruits and the coffee plant were originally used in a different way. The indigeneous people of Africa, while testing the plant, already became aware of the stimulating effect of coffee. They chewed the fruits, drank the fermented juice of the pulp, and took dried and ground beans mixed with fat on hunting raids. In Arabia, the Sheikh Hedji Omar, who was exiled from the city of Mocha, found relief in his loneliness from chewing coffee seeds and soon became famous for prescribing his coffee medicines to the sick and the weak. The first coffee infusions were probably made from the young leaves, as for tea. Later the dried coffee cherry pulp was also used; under the name kishr, it is still popular in Yemen and Somalia, and surprisingly also in Bolivia. The drinking of coffee, as we know it, is believed to originate in Aden, from where it spread. Having become popular in Grand Cairo in the 16th Century, it soon affected Southern Europe and later on Northern Europe, where coffee houses have been established since 1650 (Wellman, 1961).

Cured coffee beans contain on average 1.5 % of the alkaloid caffein, from which coffee as beverage derives its stimulatory effect. Substitutes as stimulants are tea, dry leaves of which contain up to 5 % caffein, cola nuts (2-3 % caffein) and the seeds and leaves of a few other trees and palms. Fermented cocoa beans contain about 1.8 % of theobromin, a closely allied alkaloid. Besides their stimulatory effect, coffee, tea and cocoa also have in common as beverages that they are used with boiled water, providing much liquid in the diet, largely freed from intestinal parasites.

A search for real substitutes with a taste similar to that of coffee was undertaken in Europe during periods of war, when coffee was scarce, and a range of different grains and peas were roasted, and mixed with chicory[1] and other ingredients.

1. Chicory is a hardy plant whose leaves are used for salads and whose root is cut up, dried and roasted to resemble coffee. It was first used as a substitute in Napoleonic times in France when Britain blockaded all sea traffic to France.

Plate 1. Clearing land for new plantation in coffee and cocoa zone of Ivory Ccast.

Plate 2. Nursery of Malangsari Estate, East Java, Indonesia (Rcyal Tropical Institute, Amsterdam).

Plate 3. Densely planted arabica on hillside, with banana trees. Chinchina, Colombia.

Plate 4. Mulched arabica coffee. Rwanda.

Coffee is also marketed soluble as instant coffee and to provide the mokka flavour to different kinds of food. Major by-products are the coffee pulp, used dried or fresh as fertilizer and as animal food, tannins for tanning leather and oils for the perfume industry.

1.2 Ecology

1.2.1 Climatic requirements

From the onset, a distinction has to be made between the ecological requirements of *C. arabica* and *C. canephora*. *C. arabica* is an upland species, occurring naturally as an understorey tree in forests between 1300 and 2000 m in Ethiopia at 6-9°N. The tree requires an average annual temperature of somewhere between 18-25 °C, with minimum temperatures around 13 °C and maximum temperatures not exceeding 30°C. Because of these temperatures, the altitude at which coffee is actually cultivated depends on the latitude. Between the latitudes 25°N and 25°S, the upper limits are determined by the occurrence of frost. Though on the equator arabica coffee may be found at 2500 m, in Parana (Brazil) at a latitude of 24°S, it is grown at only 100-200 m.

Being evergreen, coffee requires subsoil water at all times, except for a drier period to slow up growth, ripen the wood and initiate flower buds (Purseglove, 1977). It therefore needs a rainfall of 1500-2250 mm, well distributed through the year, but with a drier period of 2-3 months. In low rainfall areas (800-1500 mm), either irrigation is required (e.g. Kenya, Yemen) or arrangements are needed to conserve soil moisture (mulching). Too much rainfall will be tolerated easier, and in Costa Rica excellent crops are produced at an average annual rainfall of 2800 mm. However in continuously wet areas there tend to be many small flowerings which produce inferior yields.

Whereas *C. arabica* ideally requires a tropical climate, temperated by altitude and with contrasting seasons, *C. canephora* thrives well under warm equatorial climates, with average annual temperatures around 24-26 °C. It is less specific in its requirements than *C. arabica* and shows a wider range of adaptability. It is best suited to lower altitudes, in particular between 300-800 m, but in East Africa it grows well at 1300 m (Ackland, 1973). The rainfall may vary between 1000-2500 mm with an optimum of 1700 mm spread over 9-10 months, and it requires a high rate of humidity all year round. *C. canephora* does also not withstand well temperatures below 10 °C and cold winds.

Some influence on the microclimate is exercised by shade. Originally found as an understorey tree in dense forest and thought of as a heliophobe plant, coffee grows successfully without shade at higher altitudes, accompanied by mist, cloud and temperatures not exceeding 20 °C. Shade trees reduce ambient temperatures slightly and significantly reduce the temperatures of surface tissues, including leaves, flowers and buds. Slight shade might be beneficial for coffee grown with a single stem under excessive rainfall, too high or too low temperatures, less favourable soil conditions and a prolonged season with many hours of bright sunlight. It will then prevent overbearing and sustain regular yields. With intensive cultivation and optimum inputs, higher yields are obtained without shade.

1.2.2 Soil requirements

For the cultivation of coffee, the structure and texture of the soil and the topography should allow free drainage (porous) but reasonable water retention (organic material). Medium loams are ideal; heavy loams and clays are unsuitable, because of their poor aeration, and sandy soils, because they dry out too rapidly. Deep soils are considered essential as an insurance against drought and, depending on circumstances, 1.8 m is often quoted as the minimum depth. Since the coffee tree has the ability to extend and adapt its rooting system, a deep soil can compensate for a certain lack of nutrients. Preferably, there should be a high nutrient status and the soil should ideally be slightly acid (pH around 6). The soil may consist of basalt, crystalline rock and granite, as in Brazil, West Africa and India; of volcanic ash, as in Central America, Colombia and Indonesia; or of lava, as in East Africa.

The humus content of the soil is important, particularly for *C. canephora*. For new plantations, soils under forest are generally preferred, because of their friable top soil, rich in organic matter. Under these circumstances, coffee can also thrive well on slopes, as long as the treatment of the soil counteracts erosion. The topography affects most of the field operations. For economic reasons too, slopes should not be steeper than 25-30°. The newest plantations in Brazil are all established on soils that allow for mechanized harvesting.

1.2.3 Suitable zones for coffee

In the 1970s, worldwide efforts were undertaken to define agro-ecological zones and to classify major soils in the world. The FAO agro-ecological zones project bases the assessment of ecological conditions for some major crops on two main variables: (a) temperature and (b) length of the growing period. By temperature, 14 major climates are distinguished under three main headings: tropics, subtropics and temperate (FAO, 1978).

The growing period is defined as the period with an excess of precipitation over potential evapotranspiration, with an allowance for the waterholding capacity of soils. Major climatic areas and isolines for different growing periods are shown on maps 1 : 20 000 000, giving a broad index of suitability for crops by region.

The respective requirements of *C. arabica* and *C. canephora* in terms of this system are 'moderately cool' (15-20 °C) (24-hour average daily temperature) and 'warm' (over 20 °C), with a growing period of 240-330 days.

The conditions for *C. arabica* are particularly well met in Central America, in the Andes countries, some areas in Brazil, parts of East Africa and Madagascar, and in only a few areas in Asia (e.g. India and Indonesia). The conditions for *C. canephora* are found over large areas in West Africa, the lower regions of Central and South America and the Caribbean, as well as in large parts of South-East Asia.

More than 100 major soils (under 16 headings), were classified in the FAO/ UNESCO publication: 'Soil Map of the World' (SMW). In the United States Soil Taxonomy Book (ST) soils were classified into six categories and many subcatagories. Coffee is often grown on soils such as Ferralsols (SMW) or Oxisols (ST); Nitosols (SMW) and Acrisols (SMW) or Ultisols (ST), which together cover about 40 % of the total area of the tropics and about 50 % of the potential arable land. Nitosols are the most suitable, and are found, for instance, in Ethiopia and Cameroon. Ferralsols are typical of the humid tropics in Africa and are found in Uganda, Angola and Zaire, for example (Buringh, 1979). They are physically good but chemically poor. Acrisols are common in South-East Asia, but also occur in Ivory Coast and Tanzania. They are less suitable for coffee.

Table 1.3. The world production of coffee, cocoa and tea, by region over the past three decades (in 1000 t; proportion in % between brackets).

	1949/51	1959/61	1969/71	1979/81	Average annual growth rate 1950-1980
Coffee					
Africa	271 (12)	741 (18)	1278 (30)	1198 (23)	5.1
NC. America	365 (17)	569 (13)	733 (17)	918 (18)	3.1
S. America	1478 (67)	2767 (65)	1885 (44)	2425 (47)	1.8
Asia	76 (4)	184 (4)	340 (8)	576 (11)	7.0
Oceania	1 (0)	3 (0)	25 (1)	49 (1)	13.9
Total	2191 (100)	4264 (100)	4262 (100)	5168 (100)	2.9
Cocoa					
Africa	489 (66)	786 (70)	1100 (73)	999 (61)	2.3
NC. America	65 (9)	87 (8)	82 (5)	94 (6)	1.2
S. America	182 (25)	225 (20)	282 (19)	461 (28)	3.2
Asia	4 (0)	7 (1)	11 (1)	48 (3)	8.7
Oceania	4 (0)	12 (1)	31 (2)	34 (2)	7.5
Total	744 (100)	1117 (100)	1506 (100)	1636 (100)	2.6
Tea					
Africa	19 (3)	47 (5)	118 (9)	199 (11)	8.2
NC. America	1 (0)	7 (1)	34 (3)	44 (2)	13.5
S. America	601 (94)	807 (91)	1045 (83)	1462 (80)	3.0
Asia	20 (3)	36 (3)	65 (5)	128 (7)	6.4
Oceania	– –	– –	1 (0)	8 (0)	–
Total	641 (100)	897 (100)	1264 (100)	1841 (100)	3.6

Source: FAO (1952-1983)

1.2.4 Crops with similar ecological requirements

Competitive crops, as far as ecological conditions are concerned, are in the first place the other major tropical beverages: the ecological requirements for tea (*Camellia sinensis*), are close to those of *C. arabica*, though rainfall and altitude should normally be higher for tea. Cocoa (*Theobroma cacao*) has similar requirements to *C. canephora* but generally prefers a higher relative humidity. These crops also share with coffee the need for early processing. After the initial processing, the costs of transport and storage are relatively low (favourable value/weight ratio). Table 1.3 shows how the three crops are distributed over the contents: coffee is the most important in Latin America, cocoa in Africa and tea in Asia. In Section 1.3.1, the production systems of tea, coffee and cocoa are compared.

In the areas where *C. arabica* thrives well, crops such as maize and beans are usually important, as well as dairy production, whereas *C. canephora* and cocoa do have to compete in areas with cash crops such as rubber and oil palm, and food crops such as yam and cassava. Though the latter crops and cocoa are typically found within the tropical rain forest climate, *C. canephora* prefers the humid savanna climate where sugar-cane and mango trees are also found. Bananas and plantains are often grown in association with both types of coffee.

In an area such as Paraná, Brazil, which because of the danger of frost, did not appear particularly suitable for coffee, the plantations were displaced by soya bean.

1.3 Production

1.3.1 Production systems

The production process of coffee involves the cultivation and maintenance of coffee shrubs and the harvesting of the coffee berries. The pulping, fermenting and drying of the berries are dealt with under 'processing'. The coffee shrub will bear fruit for 40 or more years with greatest productivity between 5-15 years of age. In areas such as Ethiopia, West Africa and Madagascar, coffee cropping has developed as follows:

- exploitation of trees growing naturally in forests
- the cultivation of coffee trees under some remaining forest trees
- the managment of systematically established plantations without forest trees.

In other areas where coffee was not an indigeneous crop, smallholder plantations evolved from fallow farming systems as follows.

- The perennial crops planted around the house affected the rotation cycle of shifting cultivators.
- Cleared plots with arable crops were interplanted with perennial crops, which gradually began to predominate the mixed cropping system.

– As the perennial crops created more and more shade, intercropping becomes less important.

The cycle arable cultivation-fallow is in such cases replaced by the sequence arable cropping-perennial crops through which shifting systems become more stationary (Ruthenberg, 1983).

In Ivory Coast, coffee was planted after yam and cassava, and in East Africa after maize and beans. In those countries, smallholder coffee plantations are also found intercropped with such crops as bananas or plantains (East Africa, *C. arabica*) and with cocoa, bananas and pineapples (Ivory Coast, *C. canephora*).

Table 1.4. Total area, average yield and farm size of coffee producers in the world (1979-1980).

	Area (1000 ha)		Yield (kg/ha)	Size of plantation (ha)	Approx. area under small-holdings (%)
	arabica	robusta			
World	7100	2900	540		
S. America	4500	50	600	6	n.a.[2]
Brazil	2500	–	600	11	n.a.
Colombia	1100	–	700	4	n.a.
Ecuador	200	40	350	4	n.a.
Peru	150	–	600	3	90
C. America	1330	–	750	3	n.a.
Costa Rica	100	–	1100	3	n.a.
El Salvador	150	–	1100	6	60
Guatemala	240	–	700	15	20
Haiti	100	–	300	1	99
Mexico	360	–	600	4	70
Africa	1000[1]	2300	350	2	90
Angola	–	250	200	30	10
Cameroon	110	160	400	3	90
Ethiopia	650[1]	–	300	0.2	95
Ivory Coast	–	850	350	2	95
Kenya	120	–	700	0.5	70
Madagascar	20	200	400	1	90
Rwanda	30	–	700	0.1	99
Uganda	20	220	500	0.5	95
Zaire	20	210	350	5	30
Asia and Oceania	270	560	700	3	70
India	100	80	700	3	n.a.
Indonesia	50	350	600	0.5	80
Philippines	40	80	1100	4	60
Papua New Guinea	40	10	1000	5	50

1. Including 300 000 ha wild 'forest'-plantations.
2. n.a. – not applicable: no clear cut difference between large and small farm sectors.

Source: Estimates on basis of various publications

Commercial coffee was produced up to the 20th Century mainly in large estates. These started to clear new land and to plant coffee as a pure stand, with or without shade. In Brazil, coffee estates tended to shift their plantations, i.e. to replant on virgin land and to abandon run-down sites.

Coffee plantations now vary in size from large estates (over 500 ha) down to only a few shrubs and trees on small holdings. For medium-sized holdings, between 5 and 30 ha, which occur mainly in Latin American countries, the distribution between estates and private coffee plantations is becoming less clear. In about 80 % of the countries that have distinct estate and smallholder subsectors among coffee producers, the smallholdings predominate (Table 1.4). The production of coffee in most countries shows a pronounced labour peak around harvesting. So estates either engage in other activities in the remainder of the year or they rely on many seasonal labourers. Smallholders, for which coffee is usually the main cash crop, can normally well combine coffee growing with the cultivation of food crops.

1.3.2 Comparison with production systems of tea and cocoa

Generally the cultivation of tea is the most labour-intensive and so it is found predominantly in densely populated areas in Asia. However tea is advantageous in division of labour, with the harvest period in tropical areas lasting up to 12 months. This, together with the need for large scale factories near the production areas explains the tendency for tea to be grown in large plantations.

Arabica coffee is less labour-intensive with the main harvest often covering four months. As fertilizers and plant protection measures are important, labour costs are only about half of total costs. Although the better wet processing installations require a certain minimum area under coffee, this is not high (20-50 ha), and processing can be undertaken by groups of farmers (cooperatives). For the larger and less demanding robusta coffee, both labour and material inputs are reduced. Because of the inferior flavour of robusta, the simpler dry-processing method is applied. Robusta coffee is therefore suitable for peasant growers.

Cocoa is probably the least labour-intensive crop, averaging not more than 500 man-hours per hectare, against 700-2000 man-hours per hectare for coffee and more than 3000 man-hours per hectare for tea. Cocoa prefers shade and therefore smallholders usually grow it together with other tree crops (e.g. cola nuts). The first stages of cocoa processing, including fermentation and drying, can be done on farm or cooperative. As long as further processing and marketing arrangements are satisfactory, small-scale production is comparatively attractive (Andreae, 1980).

Plate 5. Flowering robusta tree. Tanzania (L.L.de Graaff).

Plate 6. Robusta tree, in production. Tanzania (L.L. de Graaff).

Plate 7. Strip-picking of robusta. Cameroon.

Plate 8. Drying of robusta cherries. Cameroon.

1.3.3 Production stages

1.3.3.1 Land preparation

Establishment includes land preparation and planting, except sometimes (e.g. in Ethiopia) where coffee berries are gathered from wild trees. The planting of coffee usually requires the raising of coffee plants in nurseries and transplanting in the field. In Brazil, however, almost all older coffee plantations were raised without nurseries.

The importance of the initial land preparation depends primarily on previous land use and the present state of vegetation. For new estate plantations to be established on virgin soil under forest cover, large operations are required to burn off trees and remove the stumps and roots, followed by ploughing and harrowing. Land previously cropped or under grass cover, as in many smallholder plantations, requires less initial land preparation but heavier fertilization. The (opportunity) cost of such land is invariably higher.

The laying out of the fields depends in the first place on the planting systems applied, the most common of which are the square, the equilaterial triangle, the avenue and the contour systems. The equilateral triangle has some advantage over square planting as it permits a fuller use of the land (20 % more trees) and allows mechanical cultivation in three directions instead of two. Avenue planting facilitates both early cropping and mechanical cultivation, but the closure of the canopy and the shading out of weeds between the rows is delayed (Webster & Wilson, 1973). The planting system should also take into account shade trees and intercrops, if envisaged.

The spacing used depends first of all on the species and variety: wider for *C. canephora* (4 m × 4 m) than for *C. arabica*: (3 m × 3 m); but also on the type and fertility of the soil, and the type of pruning adopted. In Colombia and Central America, several systems reaching densities of over 5000 trees per hectare (1.4 m × 1.4 m or 1 m × 2 m). Special high-yielding varieties are used, which need heavy rates of fertilizers. The profitability of these intensive systems depends on the price ratio between coffee and fertilizers (and other inputs). Therefore a more stable system has been developed: shade trees are allowed to develop, reducing yields and fertilizer requirements, during periods of low coffee prices, but this shade is then cut back when coffee prices increase. Shade trees mitigate climatic extremes (Section 1.2.1).

The layout of the plantations, particularly large plantations, includes that of the road system and the tracks and paths required. On slopes of 5-10°, minor conservation measures are required and on slopes of more than 10°, major soil conservation measures, such as contour planting, stubble mulching, eyebrow terraces (around each tree), interception drains along the contour lines and waterways constructed straight down the hillside, to lead away any concentration of surface water.

In the early stages of the plantations, nurse crops may be required to provide shade to the young coffee plants and cover crops to protect the bare soil between the rows against weed growth. To provide some earnings in the years before harvesting, other intercrops are also grown in smallholder coffee plantations. Bananas and plantains are often used as nurse crops but are often left when the coffee plants are mature.

1.3.3.2 Planting

The planting of coffee usually requires the raising of coffee plants in nurseries and subsequent transplanting in the field. In Brazil, however, almost all older coffee plantations were established through direct seeding. Pits for planting the coffee seedlings are generally dug well in advance of the planting season, so that the soil can weather and roots of trees crossing the pit may be killed. The top soil should be conserved, for which addition of organic manures, compost and mineral phosphate helps in the early establishment. Seedlings are obtained from the nursery in polythene bags with earth attached; roots are never bare. The coffee is transplanted to the field in the early rains at an age of 6-10 months when the plants have 5-6 pairs of leaves. For nurseries, a level or gently sloping piece of land is usually chosen near a water source. Deep fertile loam soils are preferred. Overhead shade is provided, and compound fertilizers and sometimes mulch are applied. Seeds, to be obtained from ripe berries of selected high-yielding trees are either sown densely in seedboxes at a depth of 1-2 cm, covered with sawdust or compost, and subsequently transplanted after 6 weeks, or they are directly planted in the nursery beds at a spacing of 15 to 20 cm. Land preparation and planting may require between 100 and 200 man-days per hectare.

1.3.3.3 Cultivation practices

The cultivation practices for coffee can be divided into tillage and weeding, mulching, manuring, pruning and spraying (pest and disease control). Supplementary irrigation is sometimes used (Yemen, Kenya, Costa Rica). Grasses can be harmful weeds for coffee production, the most damaging being couch grass (*Digitaria scalarum*), and stargrass or Bahama grass (*Cynodon dactylon*). Because of the fierce competition for water, tillage and clean weeding are particularly important in areas of low rainfall (applied for example in Brazil). However on sloping land, it may lead to serious erosion. In plantations without cover crops or other intercrops, the soil is sometimes dug in the first two years, and rotavated in later years. Slash weeding is usual. In West Africa, a thorough manual weeding is recommended in a circle of 1 metre around each tree. Because of ever-rising wages, chemical weed control is becoming more important.

Mulching is beneficial in coffee. It reduces weed growth, improves soil and water conservation, and adds considerable amounts of organic material and nutrients. Elephant grass or Napier grass (*Pennisetum purpureum*) is the commonest mulch. The drawbacks of mulch are that much land is needed for its production, and it incurs much labour and other costs. Besides, mulched coffee is more susceptible to frost (Brazil).

Tropical soils are generally not fertile, or only for a short period after being cleared. Few commercially grown plants extract as much nutrients from the soil as coffee.

Nitrogen is usually the major element in coffee nutrition and good yields are never maintained without regular dressings. Yield responses to the application of nitrogen are clear, with N at some 300 kg/ha, a heavy crop of 1500 kg/ha can be obtained. High rates are also needed because of the relatively easy leaching of nitrogen by heavy rainfall. Nitrogen is usually applied in 3 dressings per year, either as urea or as ammonium sulphate. If grown under shade, coffee responds less to the application of nitrogen.

Potassium is also of crucial importance in the coffee production, regulating physiological development. It is the major element in fruits and seeds. Only small amounts of phosphates are required, mainly for the promotion of flowering. After calcium and magnesium, other necessary minor elements are sulphur, iron and boron. Organic manuring, apart from that associated with mulching and cover crops, is mainly applied in mixed-farming areas at a rate of 20-30 t/ha. In other areas, coffee pulp is sometimes used. The costs of fertilizers and their application are a major element in the total costs of coffee production (Section 2.3).

Pruning can be subdivided into several activities:
- training which determines the shape of the coffee tree
- annual pruning to control growth and stimulate fruiting
- suckering, the elimination of undesired vertical shoots.

Although pruning is not applied everywhere, it is a essential practice for coffee production, in order to provide enough healthy wood for the following season's crop, to maintain the correct balance between leaf area and crop, to prevent overbearing and to reduce biennial bearing. There are also practical reasons for pruning: it facilitates picking, makes spraying more efficient and makes a less favourable microclimate for pests and diseases. The main systems of pruning, initiated at the stage of training or shaping are:

- Single-stem pruning, the simplest way of which is to retain the original seedling stem and to keep it topped from the age of 4 years to a height of about 1.80 m. Single stem pruning is often unsatisfactory for *C. canephora*.
- Multiple-stem pruning, whereby some lateral orthotropic stems are topped twice and the 3-4 stems obtained are replaced every 4-6 years by selected suckers. This method is easier and cheaper (less expertise required) than single stem pruning and is now extensively practised.
- Agobiado pruning, applied in Uganda, which is a variation of multiple-stem pruning. The main stem is bent over at an early age and pegged to the ground to allow orthotropic upright shoots to grow out.

– Candelabra pruning, used in Costa Rica, in which whereby 8 main stems are obtained on each tree to produce a candelabra effect.

Regeneration pruning is applied when the tree has deteriorated and yields have diminished. The main stem is cut (with a saw) to about 30 cm above the ground, leaving at least one upright branch.

According to circumstances, all cultivation practices require 50-120 man-days per hectare and per year of mature stands. Although to some extent depending on the distribution of rainfall and the extent of pruning practices, these practices are reasonably well spread over the year.

1.3.3.4 Harvesting

Coffee trees come into production 3-4 years after planting and are in full production at 6-8 years; some intensively cultivated special varieties, much earlier. The length of the harvesting season depends on the time span within which cherries ripen (related to climate and longitude) and also on the adopted harvesting and processing practices. When flowering and ripening occur gradually, 'selective picking' can be applied. It includes the hand-picking of all cherries that are bright red all over, firm but not hard, and leaving the unripe cherries for the next round 10-14 days later. The overripe cherries, which may become 'stinkers', are kept apart. In some countries, pickers are equipped with two baskets strapped to their waist for ripe and underripe cherries, respectively. Otherwise the cherries are graded before pulping.

The major harvesting season is usually about 4 months for *C. arabica*, and slightly longer for *C. canephora*. In the first and last months, the yields are small, and the bulk is harvested in the second and third month. In Colombia and Kenya, there is a second minor harvesting season. One picker collects about 50-100 kg of cherries per day, which corresponds to about 3 to 5 days for every 60-kg bag of green coffee. Selectively picked coffee can be pulped, i.e. processed according to the wet process, and gives a higher quality than coffee harvested by other methods, the most common of which is 'strip picking'. The pickers strip whole branches and harvest cherries at different stages of development. Harvesting proceeds much faster, which is important in areas where the ripening period is short, the plantations large and the wage rates relatively high. But the heterogeneous collection of cherries cannot be pulped and has to be processed by the dry method, resulting in lower quality and low prices. The combination of strip-picking and dry processing is common in Brazil and Ethiopia (unwashed arabica) and in several countries of Africa (mainly robusta). Harvesting contributes to more than half the total labour requirements of coffee production. In Brazil, some coffee is harvested mechanically. Complete mechanization requires an instrument to shake secondary or primary branches or the main stem itself as well as a device to collect the cherries. Prerequisites for mechanization are (almost) flat terrain, suitable soil type, well adapted varieties, appropriate spacing and specific shaping and pruning practices. It is likely to be used initially by large farmers in areas where the harvesting season is relatively short and that use dry processing.

1.3.4 Pest and disease control

The major coffee disease in the world is coffee leaf-rust (*Hemileia vastatrix*), which caused the complete collapse of arabica coffee crops in Sri Lanka and Java in the 1870s. The disease has spread throughout Asia and Africa, and has there strengthened the position of robusta coffee, of which most cultivars are resistant. The disease was first reported in Brazil in 1970 and, although much feared, does not yet have any influence on coffee production in Latin America. Control is with copper sprays at the onset of the rains.

The most serious disease in East Africa is the coffee berry disease (*Collecotrichum coffeanum*), a fungus causing brown sunken spots on the berries. Intensive fungicide spraying is required to control the disease.

Among the few serious diseases of Central and South America (except Brazil) are American leaf-spot (*Mycena citricolor*) and brown eyespot (*Cercospora coffeicola*), also requiring repeated spraying with fungicides.

Nematodes are only serious in Indonesia. Robusta coffee is hardier than arabica but it is attacked by a root-rot (*Armillaria mellea*), which is difficult to control.

Coffee pests consist mainly of sucking, eating and boring insects, which attack the tree during different stages and do require an integrated control. In several African countries the coffee bug (or antestia, *Antestiopsis* spp.) is an important pest. The coffee berry borer (*Stephanoderes hampei*) is a major pest for robusta coffee at all stages of production and even after processing, and causes the seeds to shrink and discolour. For its control, sanitary measures are required, including the removal of cherries left on the tree, and if necessary, spraying with dieldrin solutions.

Plant protection does not constitute a high proportion of labour requirements but often represents about 30-40 % of the costs of material inputs. The occurrence of major diseases can also be a major factor in the assessment of crop suitability.

1.3.5 Production levels

The total world production of about 5 million tonnes of green coffee is obtained from almost 10 million hectares, reprersenting a world yield of green coffee of about 500 kg/ha. But yields vary considerably between countries and producers: for smallholders from a mere 300 kg/ha in Ethiopia and Haiti, to 700 kg/ha in Rwanda and 1100 kg/ha in El Salvador and Costa Rica; and for estates from an average of 400 kg/ha in Zaire to more than 1000 kg/ha in Kenya, the Philippines and Papua New Guinea (Table 1.4). Exceptional yields of 3000 kg/ha and more have been obtained in Democratic Yemen, in Hawaii and on some advanced plantations in Brazil.

Because of the population pressure in Central America and Colombia, the general tendency is to decrease coffee areas, and to increase yields through closer spacing with self-shading and with higher rates of fertilizer. Though the number and efficiency of estates in Africa is declining and the total production in some countries is decreasing because of political disturbances, the smallholder coffee production in Africa has been increasing since the 1960s, and in particular in countries such as Ivory Coast, Madagascar, Cameroon and Kenya.

It must be possible with some improvements in a traditional rural environment to increase average yields for both arabica and robusta in Africa from 400-500 kg/ha to 700-800 kg/ha, but labour shortage and an emphasis on food production might restrict such a development. In Asia, total production of coffee increased considerably in the last 10-15 years, mainly in India, Indonesia, Philippines and Papua New Guinea. This has largely been the result of an increased area under coffee, but the average yield also increased from 500-700 kg/ha.

1.4 Processing

The processing of coffee consists of the following stages:

1. The initial processing of the harvested coffee cherries, which includes either the pulping, fermentation, washing and drying of the beans (wet process), or only the drying of the whole fruits (dry process).
2. The curing of the dried parchment, which includes the hulling to remove the endocarp and testa (silver skin), the polishing and the grading. After curing, 'green' beans are obtained.
3. The roasting of (blended) green beans, followed by packaging, before of after grinding.

Table 1.5. Conversion factors (%) in coffee processing. Arabica coffee is wet processed (washed) and robusta is dry processed (natural coffee).

Product	Moisture content[1]	Ratio to green coffee	
		arabica	robusta
Harvested coffee cherries	65-70	500-600[2]	450
Pulped coffee beans	–	290	–
Pulped and washed coffee beans	53	240	–
Dried coffee cherries	30-40	–	200
Dry parchment coffee	12	125	–
Green coffee	12	100	100
Roasted coffee	3	84	84
Soluble coffee	–	33	35

1. Depending on relative humidity of air.
2. Depending (among others) on altitude and shade conditions.

Plate 9. Coffee flowers. Brazil (Royal Tropical Institute, Amsterdam).

Plate 10. Selective picking of densely planted arabica. Chinchina, Colombia.

Plate 11. Measuring the harvest with 'arrobas'. Colombia (Royal Tropical Institute, Amsterdam).

Plate 12. Depulping machine, generally used in East Africa (Royal Tropical Institute, Amsterdam).

1.4.1 Initial processing

Alongside wages, the time span within which cherries ripen is a major factor in the methods of harvesting and in the first stage of processing. If flowering takes place in stages and cherries ripen gradually over a period of 3 months or more, red prime ripe berries can be picked selectively at intervals of 10-14 days. These soft ripe berries produce the highest quality and are pulpable, allowing the wet method to ferment the mucilage and resulting in a better flavour. Supply of water is a major prerequisite.

The dry method on the other hand is cheaper and is the only alternative in areas where coffee flowers at one time, resulting in a short ripening period and preventing to harvest a large portion as ripe fruit. It is also the only applicable method for green and overripe dried berries, which are too hard to be pulped. On the other hand, the juicy pulp of arabica coffee is sometimes difficult to dry.

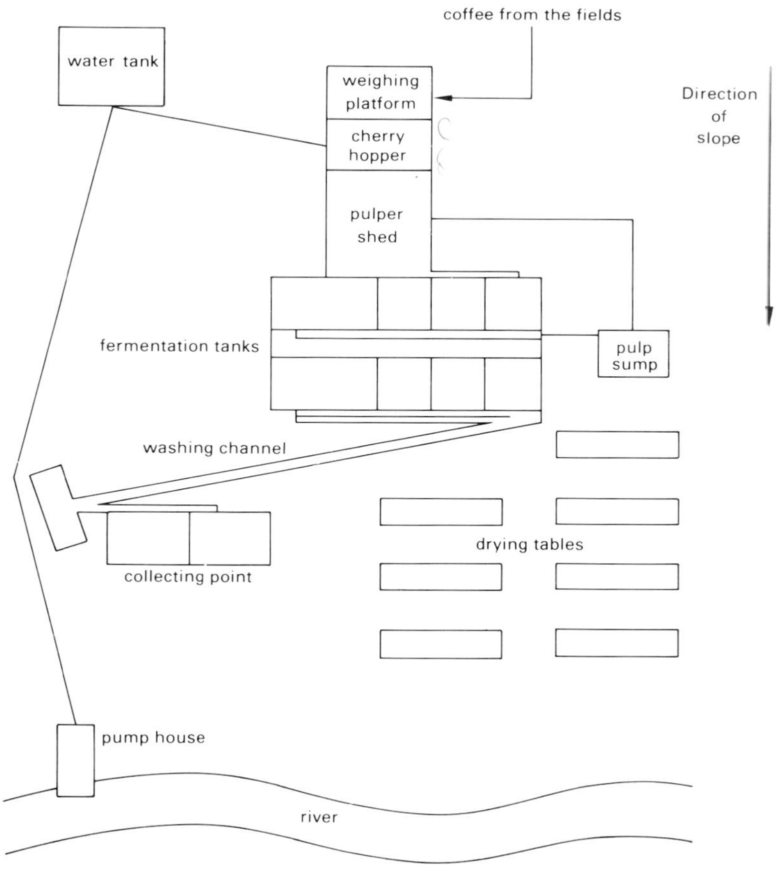

Fig. 1.2. Layout of a pulping, washing and drying factory for coffee. After Waters (1969).

1.4.1.1 Wet process (production of washed coffee)

The cherries are pulped as soon as possible after harvesting (within 24-36 h) to prevent early fermentation. The cherries are fed along with water into a pulping machine, which separates the outer pulp from the beans (parchment) and degrades the mucilage. Light and small cherries are floated off, being screened and pulped separately. The parchment then passes along grading channels into fermenting tanks (Fig. 1.2), where the sticky mucilage is broken down by naturally occurring micro-organisms and enzymes in 2-4 days. Fermentation can be accelerated by adding enzyme preparations, of special advantage in areas shortage of water. The beans are subsequently washed to get rid of the degraded mucilage and then dried in the sun (6-9 days on the drying tables) or mechanically (6-20 h). Through drying, the moisture content is reduced from about 53 % to 12 %.

The dried parchment coffee weighs only a sixth as much as the cherry from which it was produced. For this reason, factories for initial processing must be nearby. Coffee estates have their own factory or factories, whereas smallholders are either a member of a cooperative society that operates a coffee factory, or they use small hand-pulpers and basins at their homestead. Conditions are particularly favourable for washed coffee in a country such as Colombia, near the equator with little seasonal temperature change and a well distributed rainfall, resulting in a long harvest season (up to nine months). Under such conditions, a small labour force, also employed in the short off-season, can pulp and wash all the coffee with small-scale manually operated facilities.

1.4.1.2 Dry process (production of unwashed or natural coffee)

In the dry process, the fruit is allowed to remain on the tree past full ripeness and is partially dried before harvesting (moisture content decreases from an average of 65 % to about 35 %). Even if more than one pass is made, some immature fruits are still present; since separation is difficult by strip-picking as applied, a mixed product is obtained. The berries are sometimes crudely separated by flotation in water to eliminate the soft pulpable fruit, for wet processing. But the bulk is dried in the form of whole fruit, either on a sun-drying terrace in 3-4 weeks or with drying machines in 3 days (or with a combination of the two). If rain damage is likely, mechanical drying is preferred. Because of the wide range of moisture content of the cherries, drying has to be gentle. In experiments, the mucilage has also been removed chemically and mechanically. The method is more expensive but produces as good a quality coffee as the wet method.

During processing, contamination by micro-organisms (moulds, fungi, bacteria) must be avoided, since it results in inferior characteristics (e.g. 'Rioy flavour' in Brazil). Dry processing after strip picking is common in those areas where the harvesting period is short, e.g. 2 to 3 months, as in Brazil and some parts of Central America and where wage rates are relatively high. A large labour force is required for such a short period and cannot always be employed economically in the long period between harvests. Since dry processing is generally cheaper, it is more appropriate for lower-quality coffee, and is therefore usually applied for robusta coffee. Although circumstances generally favour one of the two methods of processing, a common arrangement is also to combine selective picking and wet processing with strip-picking and dry processing.

1.4.2 Curing of coffee

1.4.2.1 Hulling

Once dried and conditioned, the parchment is transported to a central factory where it is hulled, to remove parchment shell and silver skin. Slightly different hulling machines are used for wet-processed and dry-processed coffee beans, but beans are hulled in either case by creating friction between the beans lying along a rotating screw. For dry-processed coffee, a series of baffles first tears the husk off the dried cherries as these move through the machine. What is still left of the silver skin is removed by polishing and sometimes, as for Uganda robusta dry cherries with their heavy silver skins, by special machines. Section 2.4 shows the considerable differences between hulling factories in scale of operation.

1.4.2.2 Grading

The next step is the grading of the green bean, which is the first of three stages in which coffee is assessed for quality, the next two stages concerning the roasted bean and the final beverage, the liquor. Important quality factors are as follows.

- size of beans: even-sized and larger beans are preferred by consumers
- shape of beans: oval with one flat face. Pea-berries have an abnormal shape (only one seed within the coffee cherry); elephant beans (two beans grown together); and triangular beans (three beans in one berry)
- colour of beans: based on cupping test; blue or greyish blue is preferred, followed by greyish-green; brown is usually caused by processing overripe cherries, and by overfermentation
- imperfections: e.g. number of defects (such as stones, twigs, husks, sour and broken beans)
- absence of stinkers, which are beans with a bad taint, which can spoil the liquor of a good coffee.

1.4.2.3 Bagging and storage

After grading, the green coffee beans are bagged, stored and prepared for export. Normally the name of the grader-miller or exporter, and the name of the grower appear on the bag. In some countries, the name of the state or district and the port are also stated. The net weights of bags of green coffee beans differ between producing countries: e.g. Brazil, Kenya and Ivory Coast use 60-kg bags and Colombia and the Central American States 70-kg bags. Storage of dried beans in parchment in the coffee-producing countries generally results in much less deterioration than storage of the same beans hulled.

The early processing and curing together are known as green coffee technology, and is followed by roast coffee technology. Cheap labour in most of the coffee-producing countries has favoured labour-intensive production and processing techniques. Its gradual disappearance is stimulating a search for more mechanized methods of coffee production, processing, sorting, grading and transport.

1.4.2.4 Roasting

Except those using auctions, many growers already presell large consignments of green coffee beans in many countries at prices arranged by the large coffee buyers or the roasting plants in the consuming countries. Upon arrival at the plant, the green coffee beans are first cleaned, weighed and blended. Reasons for blending include the following:
- to control raw material prices
- to compensate for the heavy tasting beans of old crops with the thinner taste of new crop coffees
- to add beans with fermented tastes and off-flavour in small amounts to other consignments
- to remain flexible; the large coffee buyer prefers not to depend on one source of supply
- to offset major changes in availability and pricing of specific bean varieties in floods, droughts, diseases and frost
- to offset changes in quality of coffee beans purchased throughout the year and to obtain a certain uniformity within the range sold.

The coffee-roasting plants in the United States are situated in principal coffee-importing cities. The processing and sales are related to single-type blends. Roasts from single-brand large processors are mostly sold as-vacuum packed roast and ground coffee. The processing and marketing methods in Europe differ markedly from those in the United States mainly because of:
- the higher quality required in Europe, resulting in higher prices
- the variety of traditional and modern methods of brewing coffee, requiring a wide choice of blends, roasts and different grades of ground coffee.

During the roasting process, heat must be applied quickly and uniformly, so the beans must be moving continuously. For the first commercial installations the roasting time was half an hour per batch, with an output of about 100 bags per man-day (12 hours work). Before World War II, installations were gradually improved, resulting in a roasting time of 15 minutes, at lower temperatures (400 °C) and with an output of 800 bags per man-day.

With the introduction of the continuous roaster the productivity was doubled to 1600 bags of coffee per man-day, with one man operating two units producing 5000 kg per hour continuously (repetitive labour of loading and unloading eliminated). Beans are roasted in 5 minutes at 260 °C only.

During roasting (continuous roaster), a weight loss of 18 % occurs and fuel costs amount to 1.1 MJ per kilogram of green coffee. Roaster smoke can be eliminated by washing (water sprinkling tower) or with afterburners. The change in chemical composition of coffee beans has been shown in Section 1.1. After roasting, the coffee is further blended and then packed in cans or flexible materials.

1.4.3 *Instant coffee*

Although dry soluble coffees were sold in the United States as early as about 1900, it was in the late 1930s when Nestlé marketed instant coffee. Compared to the standard of 50 cups of brew per pound of roast coffee, recommended by the Coffee Brewing Institute in the United States, instant coffee yields 100 cups from the same pound of roast coffee.

Robusta coffees vary in flavour but are valued less than Brazilian and mild coffees in the world market. They are therefore available at a lower price and have the advantage of easily yielding larger amounts of solubles (40 %). On the other hand, the arabica mild coffees and particularly the types grown at the highest altitudes have more body and flavour (which carry through into the cup of instant coffee). There are two methods of producing soluble coffee, both involving extraction of liquor from ground roast coffee and dehydration of the liquor. Initially dehydration was always by spray-drying, but recently freeze-drying has been introduced, which conserves the taste better and results in higher yields of solubles. Freeze-drying, however, requires more energy.

The ICA (International Coffee Agreement) established 1 kg of solubles yield of instant coffee from 3 kg of green coffee beans as a basis for export quota control, corresponding to 39 % solubles yield from roast coffee. Few producers of instant coffee can compete for costs on that basis. The custom authorities Europen Community use a soluble yield of 1 kg solubles from 2.5 kg green coffee beans.

Instant coffee is a convenience food compared with brewed roast coffee. It minimizes consumer attention to preparation, eliminates cleaning after preparation, reduces product waste and does away with investment in and maintenance of a breweing device. Despite this and the cheap raw material in keeping prices down, capital investment and operating costs (skilled labour) for instant coffee manufacture are considerably higher than for roasting, grinding and packaging.

1.4.4 *Decaffeination*

Almost 10 % of the world's coffee is decaffeinated, and about half of that in European countries. The highest proportion of decaffeinated coffee consumption is found in Germany and Switzerland. Most decaffeinated coffees are sold in instant form, especially in the United States. The use of decaffeinated coffee is increasing rapidly, as is demand for robusta coffee. Though robustas have twice the caffein content of arabicas, robusta coffee has the following advantages for decaffeination.

- It gives a higher caffein yield for the processor
- The robusta coffee beans are cheaper
- So much flavour is lost in decaffeination that good quality beans are not required (Sivetz, 1979).

Caffein has been isolated in pure form since 1820. Two-thirds to three-quarters of caffein produced is used in cola beverages, the remainder being used in remedies for headache and other complaints.

1.5 Services

1.5.1 Supporting services to farmers

The importance of coffee for the economy of most producing countries, in terms of foreign-exchange earnings and taxation (Section 2.1) is to some extent reflected in the extent of services to producers, usually provided through national or quasi-governmental coffee institutes. These are generally official bodies, functioning under a coffee act and placed under the Ministry of Agriculture, or other ministry.

They are often responsible for production, trade policy and programmes, and have wide powers in implementation. Emphasis is usually on provision of technical assistance and credit for growers, price stabilization, price support, marketing and research.

In French-speaking countries in Africa, such organizations also promote other crops; in Guatemala, similar functions are performed by a private coffee organization (Anacafé). Great differences exist between the responsibilities of the coffee organizations (Table 1.6). Some are involved mainly in marketing, whereas others such as Federacafé in Colombia not only provide technical and financial assistance to farmers but are also engaged in the construction of roads, water supply and rural health. Several of the institutes are involved in processing and curing facilities.

Generally coffee seedlings, fertilizers, sprays, equipment and other facilities are made available at subsidized rates and loans are provided at low interest rates. Since coffee is becoming more a smallholder crop in most countries, much support has been given to the establishment and management of growers cooperatives, through which inputs and credit can be obtained more easily. In some countries (e.g. Kenya), these cooperatives are also responsible for the collection, initial (wet) processing and sale of coffee to the curing plant(s).

In periods of high prices on the world market, replanting and rehabilitations

Plate 13. Fermentation tanks and drying tables of Ndumberi cooperative. Kiambu, Kenya.

Plate 14. Turning parchment coffee on drying tables. Kiambu, Kenya.

Plate 15. Dried parchment coffee. Rwanda.

Plate 16. Interior of hulling installation. Rwandex factory, Rwanda.

Table 1.6. National coffee organizations and some of their functions.

	Name[2]	Functions[1]						
		Direct export role	Minimum guarantee price	Domestic supply	Processing	Infrastructure	Agric. credit	Cooperatives
Brazil	IBC	×	×	–	–	×	×	–
Colombia	Federacafé	×	×	×	×	×	×	×
Costa Rica	Oficafé	–	×	–	–	–	–	–
El Salvador	INCAFE	×	(×)	×	×	–	×	–
Haiti	IHPCAFE	–	–	–	–	–	–	×
Mexico	INMECAFE	×	×	×	–	–	×	×
Cameroon	ONCPB	×	×	–	–	–	–	–
Ivory Coast	CSSPPA	×	×	–	×	×	×	×
Kenya	CBK	–	○	○	×	×	×	×
Tanzania	CAT	–	○	○	×	×	–	×
Uganda	UCMB	×	–	–	×	–	–	–
India	ICB	×	○	○	×	–	–	–

1. ×-function exercised; ○-monopoly purchase rights.
2. IBC: Instituto Brasileiro do Cafe.
Federacafe: Federación Nacional de Cafeteros de Colombia.
Oficafé: Oficina del Café.
INCAFE: Instituto Nacional del Café (CSC).
IHPCAFE: Institut Haitíen de Promotion du Café et des denrés d'exportation.
INMECAFE: Instituto Mexicano del Café.
ONCPB: Office National de Commercialisation de Produits de Base.
CSSPPA: Caisse de Stabilisation et de Soutien des Prix et des Produits Agricoles.
CBK: Coffee Board of Kenya.
CAT: Coffee Authority of Tanzania.
UCMB: Uganda Coffee Marketing Board.
ICB: Indian Coffee Board.
Source: UNCTAD (1983)

schemes are carried out. In periods of low prices, diversification programmes are initiated. Such programmes resulted in the introduction in coffee areas of such crops as oil palm, soya bean, macademia nut and tea, with varying degrees of success (Chapter 3-10).

1.5.2 Marketing in producing countries

Though the share of domestic use in the total world supply of coffee increased from about 13 % in the early 1960s to 18 % in the late 1970s, coffee is essentially an export crop in nearly all the production countries. Exceptions are relatively small production countries such as Cuba, Malaysia, Nigeria, Panama, the Phillipines and Venezuela with less than 50 % exported.

The marketing systems in production countries generally consist of collection, assembly and processing stages, followed by export. The great diversity in coffee marketing systems between the production countries is due not only to the production systems, processing methods and to the organization of the coffee sector but also to traditional practices and to relationships between producers and trading partners.

The attempt made in Figure 1.3 to delineate marketing agencies by marketing stage is certainly incomplete, but it illustrates the many different marketing channels that exist. Large producers in a country like Guatemala customarily process, hull, grade and bag their own coffee and sell it directly to exporters. Some have

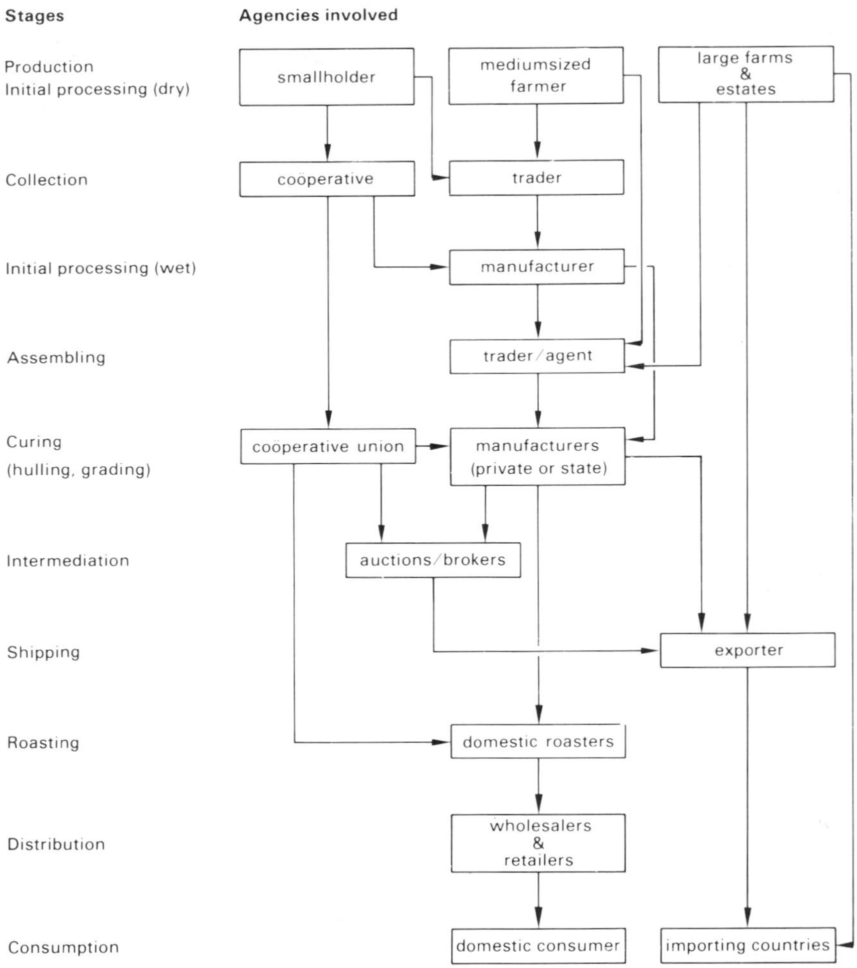

Fig. 1.3. Marketing system in a coffee-producing country.

their own brokers. At the other extreme, there are many middlemen in the marketing of smallholder coffee in Sumatra, Indonesia. There coffee changes hands regularly between collectors, agents and traders in the neighbourhood and village or district, and traders and exporters at the export harbour. In other countries (e.g. Kenya), smallholders take their cherry coffee to the cooperative pulperies, whence the parchment is taken to a major town for curing, grading and bagging, before being sold to exporters through auctions or brokers.

In only a few countries is the coffee trade completely controlled by the state, but the national coffee organizations play a major role in most countries in regulating the export and domestic marketing of coffee. In those functions, the coffee organizations can be divided into three types (UNCTAD, 1983):

- the 'marketing board' type prevalent in English-speaking coffee-producing countries (e.g. India, Kenya)
- the 'caisse de stabilization' type found in many French-speaking coffee-producing countries (e.g. Cameroon, Ivory Coast)
- the quasigovernmental coffee producers' associations and 'institutes' (e.g. Brazil, Colombia).

The marketing boards usually have a legal monopoly for the purchase of the entire coffee crop. The coffee is sold to the board, which is responsible for grading and storing it until it is resold on the export or domestic market. The board usually makes an initial payment to producers based on the average export price actually received. What margin is retained is a matter of national policy.

The 'caisse de stabilization' type of agency does not appropriate the coffee itself. It fixes the prices to be paid to the producer, the transport and marketing margins to port of shipment and the profit margins for purchasing agents and exporters, and absorbs the difference between those amounts and the export price obtained. Guaranteed grower prices are usually substantially below the world market price. The agency operates in such a way that net revenues from its activities related to some established crops (coffee, cocoa and others) encourage production of other crops.

In Latin America, coffee-marketing authorities generally establish a guaranteed minimum price at which they may purchase coffee from growers. Pricing arrangements between growers and private exporters are usually left to the free market. Growers are therefore free to choose between selling to the authority or to private firms (or to export on their own behalf), at whichever price is most advantageous. The authorities exert control over the export price and, by taxation and other means, also over the net receipts of coffee exporters. Indirectly they also influence the prices received by coffee growers.

Whatever the marketing system, the marketing channels are usually long, particularly for smallholders production, and the many institutions involved in the marketing operations contribute to considerable trade margins.

In the early history of coffee, plantations were mainly established on islands or not far from the coast (Java, Sri Lanka, Réunion, Caribbean islands). Nowadays

Table 1.7. Prices to growers and exports unit value in 1980 (Kenya and Costa Rica in 1979) (in US$ per kg green bean equivalent).

Type of coffee and country	Export unit value (f.o.b.)	Average producer price	Producer price as share of export unit value (%)
Colombian milds	(3.62)		
Colombia	3.60	1.83	51
Kenya	3.79	2.88	76
Other milds	(3.33)		
Costa Rica	3.26	2.03	62
Guatemala	3.54	2.16	61
Mexico	3.54	2.40	68
Unwashed arabica	(3.07)		
Brazil	3.03	1.51	50
Ethiopia	3.58	1.29	36
Robusta	(3.04)		
Cameroon	3.29	1.48	45
Indonesia	2.73	1.48	54
Ivory Coast	3.37	1.42	42
Uganda	3.10	1.88	61

Source: ICO (1977-1983)

coffee is generally grown more inland and as a smallholders crop in less accessible zones. Costs of collection and transport therefore became, despite the favourable weight/value ratio, relatively high, compared with other crops.

Table 1.7 shows the low share of producers prices in the value of exports in major producing countries. These are figures for one year only and hide disparities between years and between different groups of growers. But Table 1.7 shows that producers obtain a higher share of the export price in Kenya, somewhat less in Uganda and in some Central American States and a low share in other producing countries.

A major part of the trade margin for coffee is usually made up of taxes. In coffee economies such as those of El Salvador and Guatemala in 1971-1975, coffee contributed 80 % of total export tax revenues and 10-15 % of government revenues. In the boom of 1977, the export tax on coffee even made up 30 % of total government revenues. Section 2.5 shows how tax revenues from coffee relate to producer prices in some selected countries.

The taxation systems differ from country to country and are generally composed of different elements with rates fluctuating according to national production and prices on the world market. In most countries, the taxes levied at export, usually ad valorem, are the major ones, but income and other taxes are also charged at the level of production, processing and domestic marketing. A substantial part of the export tax revenues is often ploughed back into the coffee sector to finance coffee or diversification programmes and to develop the infrastructure for the coffee industry. In Colombia a major part of the tax structure is the 'reintegro' or repatriation requirement, stipulating that all coffee exporters must remit a certain sum in United States dollars for each bag of coffee exported. It acts as a minimum export price and serves to hold internal prices for coffee well below world levels.

Major marketing functions performed by the different agencies are transport, previously by rail and now mainly by road, and storage, usually as parchment coffee. Because of the quota system applied under the International Coffee Agreement, large stocks have to be kept by countries whose production continues to exceed the quotas for some years (Section 1.6). As the coffee organization of the leading exporting country concerned with the balance between production and consumption in the world, the Brazilian Coffee Institute (IBC) operates a warehousing system capable of storing 60 million bags (75 % of annual world coffee trade).

1.5.3 Research

Research activities for coffee production have been predominantly oriented towards the selection of high-yielding varieties, resistant to major diseases. General selection and breeding programmes have been carried out for arabica in Brazil (Campinas), Colombia (Chinchina), Costa Rica (Turrialba), Kenya (Ruiru) and other national coffee research stations. For robusta, research was initially concentrated in Indonesia and Zaire, but gradually switched to India, Ivory Coast (IFCC) and other African countries.

Aims include improvement of yield and vigour, better local adaptation (pest and disease resistance), less marked biennial bearing, shorter internodes for ease of harvesting, even ripening, higher proportion of clean coffee and better bean quality. Much attention has also been paid to interspecific hybrids (e.g. 'Arabusta' and 'Icatu'), combining the better flavour of arabica with robusta's resistance to major diseases, ecological requirements and less labour-intensive cultivation. Natural hybrids were also obtained between *C. arabica* and *C. liberica* ('Kawisari' and 'S26'). Cross-breeding between *C. congensis* and *C. canephora* gave a tree, 'Congusta', approaching the quality of arabica. A higher resistance to leaf-rust was often obtained.

There are high expectations from introduction of new cultivars of *C. arabica*, through wide genetic diversity collected in Ethiopia, and tried out in Cameroon, Ivory Coast and Madagascar in the past fifteen years. Some of these accessions have shown a better adaptation to lower altitudes and a higher resistance to leaf-rust (Charrier, 1982). Recently emphasis has also been paid to clonal propagations by tissue culture.

In Latin America, which until recently was free of coffee diseases, emphasis has been given to increasing yields, through applied research with short high-yielding varieties (e.g. Caturra, Catuai), higher planting densities, reduction of shade and increased use of fertilizers. In Brazil, attention is being paid to mechanical harvesting.

1.6 International trade and consumption

1.6.1 International trade

Coffee is almost exclusively produced and exported by developing countries, many of which depend heavily on it for their foreign exchange. The main markets are in the developed countries and particularly the United States and Western Europe. Originally shipped from Asia to 'coffee houses' in Europe and later mainly from Latin America (Brazil) to both Europe and the United States, coffee gradually became the major beverage in the Western World.

As a perennial crop only bearing after a few years and fully productive after 5-6 years, the price elasticity of supply is low. This is particularly so for countries whose economy depends on it and where production is mainly from smallholders (who continue to pick when the prices are low). However the demand for coffee, growing slowly in some importing countries, is stagnant in many countries. Income (expenditure) elasticity and price elasticity of demand in the major consumer countries are low (around 0.3 and 0.2 at moderate prices, respectively).

Under such circumstances, small changes in supply result in large fluctuations in price and in cyclical price movements. In years of high prices, producers tend to improve plantations and many producing countries are unable or unwilling to prevent this. In periods of low prices, most producers keep on producing coffee, though some larger production units (estates) might occasionally diversify out of coffee. Only exceptionally has coffee been destroyed and plantings been uprooted (Brazil) to keep prices reasonable (Rowe, 1963). Through this general response to world market prices, there is a tendency towards continuous stockpiling. This situation has, however, been interrupted by climatic disasters (Brazil) and political disturbances in some major production countries (e.g. Angola). On such occasions, prices rise sharply and decrease export earnings of the production countries. In the short run, these peak prices create a coffee boom in other coffee countries, but they often reduce world consumption for a long time.

The price boom of 1977 after the frost in Brazil not only reduced coffee consumption in the United States but also that in Europe, where consumers are accustomed to paying high prices for premium grades of green coffee. The impact on demand was, however, softened by high prices of tea, which is the closest substitute, other than adulterated products and soluble coffee.

The world coffee market is not a homogeneous market. In the first place, there are two main types of coffee: arabica and robusta, arabica being generally preferred by consumers because of its milder flavour. It has a lower caffein content.

Plate 17. Reception area of San Rafael integrated coffee-factory ('beneficio'). Costa Rica.

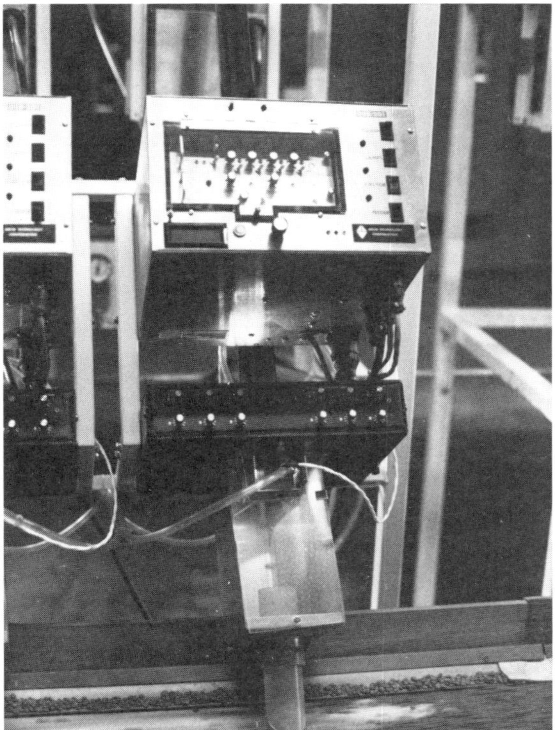

Plate 18. Electronic sorting equipment. Costa Rica.

Plate 19. Cooperative transport facilites. Bafoussam, Cameroon.

Plate 20. Factory for freeze-dried soluble coffee. Chinchina, Colombia.

According to the processing method, arabica coffees are further divided into washed or mild arabicas and unwashed arabicas. The washed arabicas are, for less obvious reasons, further classified into 'Colombian milds' and 'other milds'. There are no comparable classifications among the robustas (which are usually unwashed). Outside these main types there are certain fancy coffees such as Jamaica Blue Mountain coffee (arabica), which fetches premium prices on special markets (Japan).

Coffee is also traded in different stages of processing such as: green beans, roasted beans and soluble coffee. Of the total exports by developing countries, only 4 % is soluble coffee and less than 1 % is roasted and ground coffee. All trade statistics are converted into 'green bean equivalents', with 1 kg of soluble coffee equivalent to 3 kg of green coffee (ratio established by ICO). The share of robusta coffee has increased steadily from about 10 % in 1950 to more than 25 % in 1980. This is related in the first place to the emergence of the soluble coffee industry, for which robusta coffee is preferred because of its higher yield of instant coffee per kilogram of coffee beans. Being produced mainly in Africa, robusta coffee has also benefited from the preferential tariff of the European Community.

The four commercial types of coffee are partly real substitutes for each other and partly complementary. The roasting firms in the various consumer countries have their typical national blends, which usually include a fair amount of unwashed arabicas and a certain mixture of Colombian milds, other milds and robustas. The composition varies between countries, particularly in Europe (Section 1.6.3). Since roasters try to maintain these blends, a frost or restrictive export policy in Brazil will have a more pronounced influence on the price of unwashed arabicas then on the other types of coffee. This explains why this lower-quality coffee did fetch higher prices then the mild coffees in certain periods (Table 1.8).

Green coffee is sold on the cash markets (spot and shipment markets) and on the future markets. The best known coffee prices are those quoted on the New York spot market. The export quotas under the International Coffee Agreement are adjusted on the basis of these prices. Coffee is also traded in London and Bremen. Subtraction of transport costs to New York from the New York spot prices gives the f.o.b. prices of the various types of coffee, of most interest to coffee exporters (Table 1.8).

The quotations for Brazilian and Colombian coffee are, however, artificial, since virtually all coffee from these countries is sold under contracts with roasters in the importing countries (contracts of supply or special deals). The Brazilians grant a roaster a refund equal to the difference between the ICO indicator price for unwashed arabicas on the day of purchase and a 60-40 weighted average of the ICO indicator prices for other milds and robustas. In return for this, the roaster is obliged to purchase a specified amount of coffee, thus guaranteeing the exporting country a stable market outlet. In addition, a fidelity bonus is paid, related to the minimum of Brazilian coffee (30-60 %) used in the blend. Through these con-

Table 1.8. Price series of coffee by main commercial types and trade marks (yearly average in US$/100 kg).

	ICO-composite price[1]	Unwashed arabicas (Santos 4)	Colombian milds (Mam 5)	Other milds (Prime washed)	Robustas (Uganda Standard; Angola Ambriz)
1962/63	–	74	88	78	54
1972/73	–	131	148	127	106
1975	–	179	180	136	135
1976	–	329	348	306	282
1977	505	679	530	517	493
1978	342	364	408	359	325
1979	373	393	404	382	364
1980	332	460	394	340	325
1981	255	396	320	282	227
1982	275	317	328	306	241
Production countries		Brazil	Colombia	Mexico El Salvador, Guatemala	Uganda Angola
Some countries producing same type		Ethiopia	Kenya	India Other South American countries	Ivory Coast Indonesia

1. The ICO-indicator prices are calculated on the basis of the daily quotations collected on the New York physical market in respect of prompt shipment ex-dock New York transactions for selected other mild arabicas and robustas.

Sources: Singh et al. (1977), FAO (1975-1982)

tracts, the price actually paid for Brazilian and Colombian coffee in normal years is about the same or just above the price of other milds.

Table 1.8 illustrates clearly the extent of price fluctuations of coffee, in particular after certain events such as frost in Brazilian coffee areas. So it is not surprising that coffee is also traded by traders and speculants on future markets, the major of which are the New York Coffee, Sugar and Cocoa Exchange (coffee 'C' contracts) and London Coffee Terminal Market. Rumours about frost in the Brazilian coffee areas can easily lead to steep price increases, unless countered by provisions under the International Coffee Agreement. It is understandable that (coffee) farmers in many South and Central American countries were showing more interest in the weather forecast for the Brazilian coffee areas then for their own region.

Freight rates for coffee are established by various shipping cartels (conferences) that dominate the trade. Shipment tariffs were between US$80-120 per tonne of green coffee in 1979-1980, and total shipping costs (of which freight represents about 70 %, interest 8 % and insurance 7 %) amount to 2-3 % of coffee's landed value. Though most ships transporting coffee are owned by interests in developed countries, the Colombian and Brazilian associations Federacafé and IBC have established their own shipping lines.

Inadequate storage is still a problem in many coffee-producing countries, but the storage capacity has increased rapidly in recent years in consuming countries, and is now estimated to be 200 %. In the importing countries, coffee is handled by a variety of buyers. The green coffee is bought from producing countries by importing agents and to a lesser extent by processors. In the United States, there are about 100 green coffee importers and 200 processing companies, the first group of which handles about 70 % of total imports. These agents subsequently sell to the processing companies, to specialized roasters or to other intermediaries.

Over the past 20 years, there has been a trend towards the concentration of ownership in the production of roasted and soluble coffee in North America and Western Europe, and in many countries four firms handled more than 50 % of the trade (and in many countries all the trade for soluble coffee). Among these firms are the large diversified transnational food corporations, such as General Foods, Nestlé, Proctor and Gamble, and Consolidated Foods (including Douwe Egberts). The first two hold about 75 % of the world sales of soluble coffee.

Table 1.9 indicates the relative importance of coffee exports for some selected countries. For not less than 70 % of these countries, coffee contributes more than 25 % to the total value of exports and for 30 % of the countries coffee accounts for more than half the total exports. The level of domestic consumption of coffee per person and the size of population determine what proportion of national production can be exported, which, for most countries except the largest, is more than 80 %. The consumption of coffee per person is considerable in the major production countries (except Indonesia) but is generally low in Asia and most African countries.

1.6.2 International agreements in coffee trade

For both producers and consumers, the inherent instability in the international coffee market has long been a major problem. The international market for coffee has always been essentially oligopolistic with some large production countries, particularly Brazil playing the dominant roles. Before World War II, exports from Brazil represented almost two-thirds of the world coffee trade. During years of overproduction, it was Brazil that maintained large stocks and at times destroyed part of those to keep prices reasonable.

During World War II, the coffee trade was hampered and was further concentrated between Latin America and the United States of America. After the War, world coffee production increased rapidly and could initially be absorbed by the

Table 1.9. Domestic consumption and exports of coffee for selected producing countries (1979-1980).

	Per capita annual consumption of coffee (kg)	Coffee export		Export unit value of coffee (US$/kg)	Share of coffee in total export value (%)
		total (1000 t)	share of production (%)		
Americas					
Brazil	5.5	724	52	3.05	13
Colombia	4.5	620	83	3.55	61
Costa Rica	6.5	92	84	3.07	29
El Salvador	2.0	158	94	3.99	72
Guatemala	4.0	133	82	3.38	33
Haiti	3.0	17	53	3.76	34
Mexico	1.5	142	60	3.63	4
Africa					
Cameroon	1.0	94	92	3.14	23
Ethiopia	3.0	83	44	3.43[1]	68
Ivory Coast	2.0	225	80	3.48[1]	31
Kenya	1.0	83	89	3.57	27
Rwanda	0.1	28	99	2.57	79
Uganda	1.0	93	84	3.90	92
Asia and Oceania					
India	0.1	76	58	3.14	3
Indonesia	0.5	223	80	2.85	3
Papua New Guinea	0.1	49	99	3.39	17

1. 1979 figures.
Source: Compiled of information derived from several sources

increasing demand, particularly in Europe. Latin American and African countries increased their areas under coffee. By the end of the 1950s, world production was twice the production just after the war. In 1955, Brazil and Colombia already agreed on a policy of stockpiling. In 1957, seven Latin American countries signed an agreement in Mexico to limit their coffee export. A similar agreement was signed a year later by a group of 15 countries. The African countries did not then participate, being afraid that their increasing share of the market would be affected. The United States saw the potential danger of a collapsing coffee market for the American region and joined a 'Coffee study group'.

In the next production year 1958/59, an extremely high world production resulted in stocks equivalent to one year's consumption. Prices tumbled, particularly for robusta coffee, and in 1959 an agreement was reached, based on export quotas between the same 15 countries and the Portugese and French production areas in Africa as they then were. The outcome of the work of the study group was the successful negotiation at the United Nations of the International Coffee

Agreement 1962. The International Coffee Organization, administering that and following agreements, has the backing of virtually all of the world's coffee exporting countries (47) and a large proportion of the world's coffee importing countries (24).

A second five-year agreement went into operation in 1968. These first two agreements contained provisions for the use of a quota system by which supplies of coffee in excess of consumer requirements were withheld from the market. Under other provisions of these agreements, production and diversification policies were initiated with the objective of limiting coffee supplies. On the other hand, promotion activities were instituted with the objective of increasing consumption. These two agreements contributed to relatively stable prices in the period 1963 to 1975. Stocks diminished, in particular in Brazil. That country developed a soluble coffee industry, tax-free exports of which were to compete with the United States instant coffee industry. This situation caused a special trade arrangement for coffee between the United States and Brazil.

Meanwhile the world coffee market had become less transparent through the increasing use of 'special deals' between some major production countries and large roasting firms in consumption countries. Under the influence of the devaluation of the dollar in 1972/73, the economic clauses of the second agreement were suspended between 1973 and 1976. The four major production countries, Brazil, Colombia, Angola and Ivory Coast, formed the 'Geneva Group' before a new International Coffee Agreement was reached in 1976.

A year after the serious frost in Brazil, the situation was different from that before the Agreements of 1962 and 1968, and doubts existed about the adequacy of supplies to meet demand in the immediate future. New provisions were added to stabilize prices and to adopt constructive production policies:

- basic and flexible quotas and their suspension if prices rose above certain levels
- the strengthening of ICO to act as a centre for studies and economic research on the production, distribution and consumption of coffee
- special provisions for small producing countries.

ICO, with its headquarters in London and functioning through the International

Table 1.10. ICO-votes of major exporting and importing countries.

	1983	1976		1983	1976
Brazil	228	336	United States	294	392
Colombia	166	114	West Germany	127	104
Ivory Coast	57	49	France	89	87
Indonesia	57	26	Italy	60	–
El Salvador	49	35	Japan	53	37
Uganda	42	32	Netherlands	45	47
Mexico	42	32	United Kingdom	43	51
Guatemala	39	33	Canada	32	32

Coffee Council, the Executive Board and the Executive Director, has been assigned several new duties since 1976. From 1976 to 1980, this included the operation of a system by which shipments were covered by a Certificate of Origin to eliminate 'tourist coffee' (coffee traded through third countries); after 1980, the operation of an export quota system; and an obligatory system of controls. It also includes the annual verification of stocks carried over in all exporting member countries, the gathering and dissemination of statistical information (Quarterly Bulletins) and a computerized data base (Coffeeline). The two groups of exporting and importing countries each have a total of 1000 votes in ICO; though each of these countries has a basic vote of 4 and 5, most votes are redistributed according to share in the trade (Table 1.10). The latest Internatioal Coffee Agreement, implemented on 1 October 1983, maintains and strengthens the provisons of the 1976 Agreement.

1.6.3 Worldwide consumption of coffee

Only a small proportion of the world's population drinks coffee regularly and two-thirds of the world's coffee is consumed in the United States, Europe and Japan: countries with a high standard of living. In many of these countries, taxes and duties on coffee were fairly high in the past, making it an expensive consumer

Table 1.11. Per capita consumption, total disappearance and import and retail prices of coffee in major consuming countries.

Major consuming countries	Per capita consumption of coffee (kg/yr)	Disappearance of coffee products (1000 t)		Import price (cif) in 1979	Retail price in 1979	Margin as share of retail price (%)
		1979	1980	(US$/kg)		
United States	5	1264	1100	3.34	5.07	34
EEC		(1305)	(1312)			
West Germany	7	364	397	3.64	10.00	64
France	6	298	304	3.45	7.16	52
Italy	4	210	225	3.58	7.74	54
Netherlands	8	129	107	3.64	6.11	40
United Kingdom	2	133	125	3.13	6.57	52
Other ICO-members		(779)	(830)			
Canada	4	107	103	3.68	5.44	32
Japan	2	142	198	3.63	10.88	67
Spain	3	93	93	3.28	7.18	54
Sweden	12	113	93	3.71	6.46	43

Source: UNCTAD (1983)

commodity. Tariffs are now (1983) low (in the United States zero; in the European Community 7 % for unroasted, 15 % for roasted and 18 % for soluble coffee) but other barriers and consumption taxes are still prevalent in some countries.

The United States market differs basically from that in Europe, because of its high proportion of instant coffees (20 %), the use of milk and the preparation of light brews. This results in the import and blending of a larger proportion of robusta and unwashed arabicas and a lower proporition of quality beans. Similar drinking habits are found in other Anglo-Saxon countries where tea is usually a more important drink.

The blending, preparation and use of coffee is different for each European country. Because of former colonial affiliations and preferential treatment in the European Community, coffee has a strong robusta flavour in Western Europe, particularly in France, but also in Portugal, Belgium and the Netherlands. By contrast, the Scandinavian countries and Italy use a large proportion of Brazilian coffees and these countries are unlikely to consume large amounts of instant coffee. Internal taxes are high in West Germany, which mainly imports better-quality coffees (Colombian milds). Besides differences in blending, there are differences in roasting (e.g. dark roasts in Italy) and in preparation (espresso, mexican-sugared coffee).

Information on consumption of coffee per person in selected producing and major consuming countries is given in the Tables 1.9 and 1.11, respectively. Table 1.11 also shows the relative importance and recent trend of consumption in importing countries and regions. Consumption decreased in the period 1976-1980 in the United States, remained about the same for the countries of the European Community and increased in Japan.

Retail prices and trade margins (including taxes) were (in 1979) extremely high in Japan and West Germany, and low in the United States and Canada.

References

Acland, J.D., 1973. East African crops. Longman, London.

Andreae, B., 1980. The economics of tropical agriculture. Commonwealth Agricultural Bureaux, Slough.

Buringh, P., 1979. Introduction to the study of soils in tropical and subtropical regions. Pudoc, Wageningen.

Charrier, A., 1982. L'amélioration génétique des cafés. La Recherche 13 (136): 1006-1016. Paris.

Coste, R., 1968. Le caféier. Ed. G.P. Maisonneuve et Larose, Paris.

FAO, 1952-1983. Production Yearbook. FAO, Rome.

FAO, 1975-1982. Monthly Bulletin of Statistics. FAO, Rome.

FAO, 1978. Report on the agro-ecological zoning project. Volume 1-4. FAO, Rome.

Haarer, A.E., 1962. Modern coffee production. Leonard Hill (Books) Ltd., London.

ICO, 1977-1983. Quarterly Statistical Bulletin on Coffee. ICO, London.

IFCC, 1964. Le calendrier agricole, du planteur Ivoirien de café et cacao. Côte d'Ivoire.

Krug, C.A. & de Poerck, R.A., 1968. World coffee survey. FAO Agricultural Studies No. 76. FAO, Rome.
Marshall, C.F., 1983. The world coffee trade. Woodhead-Faulkner, Cambridge.
Purseglove, J.W., 1977. Tropical crops; Dicotyledons. Longman, Essex.
Rothfos, B., 1980. Coffee production.
Rowe, J.W.F., 1963. The world's coffee. Her Majesty's Stationery Office, London.
Ruthenberg, H., 1983. Farming systems in the tropics. Clarendon Press, Oxford.
Singh, S., Vries, J. de, Hulley, J.C.L. & Young, P.,1977. Coffee, tea and cacao; Marketing prospects and development lending. Worldbank Staff Occasional Papers No. 22, Washington.
Sivetz, M.Ch.E. & Desrosier, N.W., 1979. Coffee technology. AVI Publishing Company INC, Westport, Connecticut.
UNCTAD, 1983. Marketing and processing of coffee; areas for international cooperation. UNCTAD, Geneva.
Webster, C.C. & P.N. Wilson, 1973. Agriculture in the tropics. Longman, London.
Wellman, F.L., 1961. Coffee; Botany, cultivation and utilization. Leonard Hill, London.

2 Comparative analysis

2.1 Framework

2.1.1 Scope of analysis

This monograph is concerned with the economics of coffee in producing countries and with the supply of coffee to the world market, since two-thirds of the world's coffee production is exported.

Analysis of the economics of coffee in producing countries requires a comprehensive approach, in which the commodity is followed through the production, processing and marketing stages to its export or domestic consumption. Other aspects than economics, such as organizational structure and policy matters, are also considered. The object of analysis is thus the coffee sector, defined as those parts of the agricultural, industrial, trade and services sectors that deal with coffee and products derived from it. Ecology is added to the scope of analysis, since ecological conditions in producing countries are the basis for coffee cultivation, as well as the starting point for considering alternative crops or alternative agricultural enterprises.

The coffee sector has been studied in eight producing countries and these case studies provide the elements for the comparative analysis, resulting in conclusions about basic factors, production, processing, marketing, services and policies, together with relevant economic parameters, such as costs, income and income distribution, employment and earnings of foreign exchange.

In some of the major coffee-producing countries, such as Brazil, Mexico, India and Indonesia, there is a considerable demand for coffee. However most coffee is exported and it is the major export earner for many producing countries. The world's coffee market is now largely controlled by quotas and regulations under the International Coffee Agreement. This agreement limits the scope for individual countries to increase exports and it also largely determines the unit value of exports. Section 2.5.1 considers the main factors that influence export prices. The comparative analysis emphasizes factors in production and sales, and cost elements in production, processing and marketing. The producing countries each have a different history and ecology, and different resources (Section 2.2), which greatly influence the production systems adopted and hence the production costs (Section 2.3). Various factors influence methods and costs of processing and marketing, which also partly depend on the degree of government intervention and the organization of the coffee sector (Section [illegible]). Whereas the position in the world market and the level of production determine the gross earnings of the coffee sector, the various cost elements, the domestic pricing policies and the rates of taxation influence net earnings and their distribution among the participants in the coffee sector (Section 2.5). The main conclusions are presented in Section 2.6.

2.1.2 Selection of countries

Coffee is grown by many countries in different parts of the world (Figure 2.1; Table 1.4). This comparative study includes the main producing countries, and a careful selection of some other typical producing countries.

The main types of coffee distinguished by the International Coffee Agreements for the allocation of basic quotas to the major producing countries are as follows (the main producers in parenthesis):
- unwashed arabicas (Brazil and Ethiopia)
- Colombian milds (Colombia, Kenya and Tanzania)
- other milds (Central America and India)
- robustas (Ivory Coast, Indonesia, Uganda, the Philippines and Cameroon).

Some countries produce both arabica and robusta coffee, but one of the two is usually the main export (e.g. Brazil, arabica; Indonesia, robusta). Ecuador, Cameroon, Tanzania and India produce sizeable amounts of both types. Ivory Coast is now seeking to introduce its arabusta coffee (hybrid between arabica and robusta) as a separate type. Among the minor producing countries, under ICA generally exempted from basic quotas, are several American and African arabica-producers, the most prominent of which are Venezuela, Haiti, Rwanda and Burundi, and some African and Asian robusta-producing countries.

Fig. 2.1. Main coffee areas in the world.

In the selection of countries for study, it was considered essential to include
- the principal producing countries in the three continents of America, Africa and Asia
- representative countries for the main types of coffee
- a country producing both arabica and robusta coffee
- one or more countries with both an estate and a smallholder coffee sector
- a minor coffee-producing country.

Of the 22 countries listed in Table 1.4, the following 8 countries were selected for the case studies (with their ranking in the world (1981/83) and their main features).

Brazil, 1: principal producer of unwashed arabica
Colombia, 2: principal producer of Colombian milds
Ivory Coast, 3: principal producer of Africa; 1st robusta-producer
Indonesia, 4: principal producer of Asia; 2nd robusta-producer
Costa Rica, 12: representative of Central American other milds
Cameroon, 13: producer of both arabica and robusta
Kenya, 15: African arabica-producer, with estates & smallholders
Rwanda, 28: minor producing country, with semiwashed (mild) arabica.

In the case studies (Chapters 3-10), the 5 arabica-producing countries are presented first, and geographically the sequence Latin America, Africa, Asia is followed. The 8 countries together represent about 60 % of the world production of coffee. Table 2.1 present some basic data for these countries and illustrates the role of coffee in the national economies. It shows that the group of 8 countries comprises some low income countries (Rwanda and Kenya); several lower mid-

Fig. 2.2. Countries selected for case studies.

Table 2.1. Coffee in the national economies of the selected countries (1982).

	Bra.	Col.	Cos. R.	Ken.	Rwa.	Cam.	Ivo. C.	Ind.
General indicators								
GNP per capita (US$)	2240	1460	1430	390	260	890	950	580
population density (persons/km^2)	15	26	45	32	220	20	28	84
energy imports (share (%) of merchandise export 1981)	52	25	21	63	.	13	21	8
Role of agriculture (%)								
share of GDP	13	26	25	33	46	27	26	26
share of total employment (1980)	30	26	29	78	91	83	79	58
share of total exports[1] (1981)	45	70	67	52	.	64	82	13
Role of coffee (%)								
share of GDP	1	7	5	4	5	5	6	1
share of total employment	2	5	5	3	2	4	7	1
share of total exports	10	51	27	24	63	17	22	2

1. Under category: primary commodities other than fuel and minerals.
Sources: World Bank (1984), country studies (chapter 3-10)

dle-income countries, of which one major mineral oil exporter (Indonesia) and one upper middle-income country (Brazil). As it neatly shows, the role of agriculture becomes less important with higher the income per person; agricultural exports, in particular coffee, are by far the major export for most of these countries, except for the OPEC country Indonesia.

2.2 Basic factors in coffee cultivation

2.2.1 Historical development and ecology

In the course of history and through experience and research, more knowledge has become available about specific requirements for the cultivation of coffee. Found growing wild in forest areas in African countries, such as Ethiopa (arabica), Uganda and Zaire, (robusta) coffee was gradually introduced into other countries by traders, missionaries and Government officials during the 18th and 19th Centuries. The colonial administration in many tropical countries considered coffee an excellent export crop, that could raise considerable tax revenues. In some countries, it expanded after the collapse of the world market in other commodities (indigo and tobacco). For Latin America, in particular Brazil, this coincided with the emergence of a major new market: the newly independent United States of America.

Arabica coffee was first introduced in areas close to export harbours and hence at low altitudes (Réunion, Caribbean islands, Brazil), where there were also sugar-cane plantations, or the hilly areas surrounding such plantations. However settlers soon realized that arabica coffee thrived better at temperatures of 17 to 20

°C, which in the tropical belt are usually found at an altitude of at least 1000 m. Since they themselves preferred such climatic conditions, the settlers (in such countries, such as Brazil, Colombia, Costa Rica, Indonesia and Cameroon) started to move towards the interior of these tropical countries, and constructed railways and roads to improve communication between the interior and the harbour towns. This in turn stimulated the production and marketing of coffee, which by 1850 had become a major beverage in the Western World.

With the settlers and the traders (coffee was a business), the work force also moved towards higher altitude zones in the interior, which gradually became more densely populated. There coffee, tea and sometimes cotton became the main estate crops. Estates were also involved in livestock production. Tea had the advantage of offering regular employment throughout the year, whereas picking of cotton in the drier zones of certain countries offered work during the slack period of the coffee crop. After abolition of slavery and the dissolution of the different types of labour contracts, estate coffee did not increase any further and declined in several countries. Production on estates was gradually overtaken by production by peasants, who had become well acquainted with coffee, while working on the plantations (Section 2.3.1).

A new processing method, 'wet processing', was discovered in the 19th Century and applied in Indonesia and later also in Central America and Colombia. For washing and fermentation, a considerable amount of water is required and, for the subsequent drying, some area of land. The larger farms could usually find suitable sites for this processing, but the small and very small farms needed to form cooperatives to perform these processing activities (Section 2.4). Now large and medium-sized farms still play a major role in Latin America, and smallholders are the main coffee producers in Africa.

Development of robusta production came later and originated mainly from Zaire. In Indonesia, robusta displaced arabica coffee when the latter type was largely destroyed by leaf-rust around 1900. Robusta was a major estate crop in Angola and Zaire, but proved to be an excellent smallholder crop in many other countries because of its simple production and processing. The two types are, however, not usually found in the same areas, since robusta requires somewhat higher temperatures and is therefore generally grown below an altitude of 1000 m.

Table 2.2 gives climatic data for some stations in the countries selected. Within the belt from 15°S to 15°N of the Equator, arabica and robusta coffee are found above and below 1000 m respectively. This corresponds with average daily mean temperatures of about 19-20 °C for arabica and around 24 °C for robusta coffee and a range of extreme minimum and maximum temperatures of 5-32 °C and 14-35 °C, respectively. In Brazil, arabica coffee is grown at a higher latitude, and a lower altitude, resulting in a similar daily mean temperature of 20 °C, but with a wider range between the extreme minimum (−2 °C) and maximum (37 °C) temperatures. So those production areas are less suitable for coffee production, as

Table 2.2. Climatic data for some important stations in the selected countries.

	Latitude	Altitude (m)	Temperatures[1] (°C)			Rainfall (mm)	
			daily mean	extremes		annual	monthly in dry period(s)[2]
				max.	min.		
Arabica areas							
Brazil							
Campinas	23°S	663	20	37	−2	1393	47 (5)
Colombia							
Chinchina	5°N	1310	21	33	6	2530	
Costa Rica							
San José	10°N	1120	20	33	9	1944	21 (5)
Kenya							
Kiambu	1°S	1767	18	30	2	1014	35 (4); 46 (2)
Rwanda							
Rubona	2°S	1706	19	28	10	1139	30 (4)
Cameroon							
Dshang	5°N	1398	20	32	9	1919	31 (4)
Robusta areas							
Cameroon							
Yaoundé	4°N	759	23	35	12	1587	37 (3); 57 (1)
Ivory Coast							
Adzopé	6°N	150	26	36	15	1248	34 (3)
Indonesia							
G. Menggang	5°N	550	23	32	16	2241	

1. Data on temperatures in some cases derived from nearby stations: Colombia-Medellin; Kenya-Nairobi; Ivory Coast-Abidjan.
2. Dry periods include all months with less than 60 mm rainfall; in brackets number of dry months.
Source: Landberg (1976)

Brazil has found out by experience (damage by frost).

The average annual rainfall for the respective zones in the selected countries ranges from a mere 1000 mm in Kenya to more than 2000 mm in Colombia and Indonesia. In these two countries, the rainfall is well distributed, without a distinct dry season, so that the flowering and harvesting season are prolonged. Costa Rica, Kenya and Rwanda have rather long and pronounced dry seasons, and supplementary irrigation and mulching are therefore necessary. In some areas in Kenya, there is a bimodal rainfall pattern, resulting in a second harvesting period.

When densely planted or planted along the contours, coffee can be grown on steep slopes, where other crops would cause much more erosion. It is therefore often grown on slopes in hilly areas.

2.2.2 *Production factors*

Although material inputs have become increasingly important, especially for intensively cultivated arabica coffee, the most prominent production factors in coffee production are usually land and labour (Section 2.3). These two factors are combined in different ways according to their relative scarcity. In a country like Brazil, with a population density of 15 per square kilometre, the production of 1 t of green coffee requires on average 1.7 ha of land and only 0.5 man-equivalent (150 man-days), whereas, in the densely populated country of Rwanda (220 per square kilometre), it requires 1.4 ha and 1.9 man-equivalent. Table 2.3 shows how much land, labour and nitrogen fertilizers are used by farmers in the selected countries, for production of 1 t of green coffee. In densely populated zones (e.g. Kenya and Rwanda) more labour is used, whereas in countries with a more favourable ratio of men to (agricultural) land (e.g. Brazil, Ivory Coast and Cameroon) much land and less labour is used in coffee production (low yields). In Indonesia, coffee is mainly produced in the Island of Sumatra, which is less densely populated than Java.

Tables 2.4 and 2.5 show how much use is made of land and labour in the selected countries for the production of coffee and, given the number of farms with coffee how much land and labour the average farmer devotes to coffee. In Colombia and Costa Rica, at least 18 % of the cultivated area is planted with coffee, and a slightly smaller proportion of the agricultural work force is engaged in the cultivation of coffee. In Ivory Coast, coffee occupies about a quarter of the agricultural land but coffee growing is not labour-intensive.

Although coffee is already grown in areas of marginal suitability in most producer countries, there is still some land available for expansion of coffee in Brazil, Indonesia and Cameroon (robusta only). But in the other countries included in the case studies, there is hardly any land available to expand coffee production,

Table 2.3. The average combination of inputs of land, labour and fertilizers for the production of 1 t green coffee in the selected countries.

	Land (ha)	Labour (man-years)	Fertilizers (kg N)
Brazil	1.7	0.5	150
Colombia	1.3	0.9	100
Costa Rica	0.8	0.6	100
Kenya	1.2	1.7	100
Rwanda	1.4	1.9	25
Cameroon	2.8	1.3	60
Ivory Coast	3.3	1.0	15
Indonesia	2.0	1.0	80

Source: Country Studies (Chapter 3-10)

Table 2.4. The factor land in coffee production in the selected countries (1982).

	Total land area (million ha)	Area under arable & permanent crops (million ha)	Proportion area cropped (%)	Area under coffee (1000 ha)	Proportion of area cultivated (%)	Number of coffee farms (× 1000)	Average coffee planting (ha)
Brazil	845.7	73.3	9	3200	4	300	10.7
Colombia	103.9	5.7	5	1010	18	300	3.4
Costa Rica	5.1	0.5	10	90	18	33	2.7
Kenya	56.9	2.3	4	130	6	320	0.4
Rwanda	2.5	1.0	39	35	3	500	0.1
Cameroon	46.9	6.9	15	370	5	330	1.1
Ivory Coast	31.8	4.5*	14	1050	23	350	3.0
Indonesia	181.2	19.5	11	670	3	650	1.0

Sources: FAO (1982), Country studies (Chapter 3-10)

Table 2.5. The factor labour in coffee production in the selected countries (1982).

	Total population ($\times 10^6$)	Labourforce in agriculture ($\times 10^6$)	Proportion of total labourforce in agric. (%)	Labour required in coffee (1000 man-years)	Proportion agric. labour engaged in coffee (%)	Number of coffee farms (× 1000)	Labour input for coffee per farm (m/yr)
Brazil	126.8	20.9	30	1,000	5	300	3.3
Colombia	27.0	4.2	26	700	17	300	2.3
Costa Rica	2.3	0.4	29	60	15	33	1.8
Kenya	18.1	6.6	78	200	3	320	0.6
Rwanda	5.5	2.6	91	45	2	500	0.1
Cameroon	9.3	4.2	83	170	4	330	0.5
Ivory Coast	8.9	3.7	79	290	8	350	0.8
Indonesia	152.6	50.4	58	360	1	650	0.6

Sources: World Bank (1984), Country studies (Chapter 3-10)

which already faces heavy competition from food crops.

The average size of coffee plantations is largest in Brazil, followed by those in Colombia, Ivory Coast and Costa Rica. In Kenya, large estates still exist but more than 60 % of the coffee is produced by smallholders on plots of about 0.2 ha. In Rwanda, coffee plots are still smaller (Table 2.4).

Coffee farms in the three Latin American countries require about 2 to 3 man-equivalents and a large proportion of these farms need to hire labour. In the African countries and Indonesia by contrast, much of the work is done by the farmer and his family. Much of the cultivation and harvesting in many countries is undertaken by the men, with the women tending the food crops. But manual sorting during processing is usually done by women.

Particularly in countries with an unfavourable man-land ratio and where coffee competes with food crops, much attention has been given in the 1970s to the breeding and introduction of compact varieties, which at high planting densities and with high levels of fertilization give much higher yields. Such varieties have already been introduced largely in Costa Rica and Colombia (Caturra) and will soon be released in Kenya (Catimor).

2.3 Production systems and production costs

2.3.1 Production systems

The ecological conditions in the production areas, the man-land ratio and the type of planting material largely determine the farming and production systems. Specialized coffee farms and estates still exist in countries like Brazil, Colombia, Kenya and Indonesia, but most holdings with coffee are small or medium-sized family farms with other agricultural activities. In Latin America, these farms are often 10-100 ha, in which range tractor inputs are used for land preparation and hired labour for harvesting. In Africa by contrast, most coffee farms are below 2-3 ha, since labour is scarce and power inputs are seldom used (Table 2.6). Usually the lower the man-land ratio, the lower the planting density.

Rainfall largely determines cultivation practices. In areas of low rainfall, either mulching (by smallholders) or supplementary irrigation (on large farms and estates) is required. In areas with excessive rainfall, weed control is a major activity, requiring considerable labour inputs or herbicides. Under such conditions, farmers might to some extent avoid these additional labour inputs for their coffee by interplanting it with annual crops or by growing the coffee under shade to reduce weed growth.

The rainfall distribution largely influences the harvesting season(s). One distinct dry period normally leads to a harvesting period of no more than 3 months. Under other conditions, harvesting is extended over a longer period or split into two seasons. Since the labour requirements for harvesting usually constitute about half the total labour input (Table 2.7), other enterprises and activities must compensate for underemployment for the rest of the year.

This is a major reason why both small and large coffee farms are usually diversified. Small farms usually grow the local staple food crop, which is maize in most of the arabica production areas. In robusta zones, plantains, cassava and tubers are the major food crops, and sometimes rice (Section 2.3.3). For estates, diversification is more difficult and is mainly confined to livestock enterprises or large-scale maize production. Since there is hardly any scope for mechanization (except in

Table 2.6. Main coffee production systems in the selected countries.

	Size farm with coffee (ha)	Area under coffee (ha)	Plant density (plants/ ha)	Cultivation methods and intercrops[1]	Other major farm enterprises
Brazil	60.0	11.0	1200	Pure stand	Livestock, maize, soya beans
Colombia	15.0	3.5	2700	P/with or without shade	Livestock, sugar-cane, maize
Costa Rica	30.0	3.0	3400	P/(un)shaded or M (maize & beans)	Dairying, sugar-cane
Kenya					
estates	70.0	55.0	1500	P/suppl. irrigation	Maize, livestock
smallholders	1.5	0.2	1300	P/mulched	Maize, beans, dairying
Rwanda	1.0	0.1	1700	P/mulched	Bananas, beans, sorghum
Cameroon					
arabica	1.2	0.8	1200	M (with maize, veget.)	Bananas
robusta	2.5	1.2	1300	Partly M (with cocoa)	Plantains, cassava
Ivory Coast	5.5	3.0	900	P or M/shaded (cocoa, cola)	Plantains, tubers
Indonesia					
smallholders	2.0	0.6	1400	Forest/gardens (with pepper, cloves)	Paddy, bananas coconuts

1. P = pure stand; M = stand.
Source: Country studies (Chapter 3-10)

Table 2.7. Distribution of labour inputs (in man-days) by activity in the selected countries.

Activities	Bra.	Col.	Cos. R.	Kenya est.	Kenya smallh.	Rwa.	Cameroon arab.	Cameroon rob.	Ivo. C.	Ind.
Weeding	20	40	20	35	25	40	20	40	30	40
Fertilization & spraying	8	10	10	40	15	20	17	10	1	10
Mulching	5[1]	–	–	20	25	40	–	–	–	–
Pruning & shade regulation	–	10	20	50	10	5	8	10	1	10
Harvesting	35	67	85	150	120	120	31	35	32	50
Processing & drying[2]	5	12	–	55	–	35	9	5	2	4
Other	2	6	15	50	25	15	5	10	4	6
Total	75	145	150	400	220	275	90	110	70	120

1. Sweeping.
2. In Costa Rica and Kenya (smallholders) processing is not carried out on the farm.
Source: Country studies (Chapter 3-10)

Brazil), the coffee estates are faced with the problem of hiring a considerable labour force during harvesting. This labour force is not always available (smallholders having their own coffee) or only at high cost. This factor and a lack of management capabilities explain the gradual decline in the estate sector.

In the selected countries, the area under coffee generally constitutes about 10-50 % of the total farm size (Table 2.6). The coffee estates in Kenya are specialized farms. The producers of arabica coffee in Cameroon also have a large part of their (small) farms under coffee but most is interplanted with food crops. Planting densities are low in Ivory Coast and Brazil where extensive cultivation practices are used, whereas they are high in Costa Rica and Colombia, illustrating the extent to which compact varieties have been adopted. With low rainfall, shade is disadvantageous but, in other conditions, it usually extends productive life, reduces weed control and provides some firewood. Unshaded coffee normally gives higher, early yields and responds better to fertilizers.

2.3.2 Production costs

The establishment costs of coffee plantings depend largely on the type of planting material and the planting density. For compact varieties, the establishment period is usually 3 years against 4 or even 5 years for traditional varieties. Table 2.8 compares the total establishment costs for the countries, sometimes specified for traditional and modern plantations. Labour is the major cost. Only in Brazil is extensive use made of tractor. For modern densely planted arabica, establishment costs are around US$3000 per hectare and for a modern robusta plantation about US$1000 per hectare.

Table 2.8. The establishment costs per hectare of coffee plantings in the selected countries.

	Plant density (plants/ha)	Establishment period (years)	Establishment costs (3-4 year period)	
			total (US$/ha)	labour inputs (man-days/ha)
Brazil	1000	3	900	140
Colombia (trad.)	1500	4	1500	240
(modern)	4500	3	3400	460
Costa Rica	3400	3	3000	.
Kenya (present)	1300	4	1600	510
(future)	4000	3	2800	580
Rwanda	1700	4	1150	730
Cameroon	1300	4	650	280
Ivory Coast (trad.)	900	5	500	180
(modern)	1400	4	1000	300
Indonesia	1400	4	750	300

Source: Country studies (Chapter 3-10)

Table 2.9. The production costs per hectare of green coffee in the selected countries (1982)

	Costs of labour			Costs of material inputs	Over-head & esta-blishment costs	Total costs	Yield	Cost per kg coffee
	(man-days)	wage per worker (US$/day)	costs (US$)	(US$)	(US$)	(US$/ha)	(kg/ha)	(US$/kg)
Brazil	75	3.0	220	220	280	720	600	1.20
Colombia	150	4.0	620	200	520	1340	800	1.70
Costa Rica	150	2.4	360	380	580	1320	1200	1.10
Kenya								
estates	400	1.5	600	750	800	2150	1100	1.95
smallholders	220	1.2	270	220	280	770	600	1.30
Rwanda	275	1.2	340	190	290	820	700	1.20
Cameroon								
arabica	90	2.0	180	70	90	340	200	1.70
robusta	110	2.0	220	50	100	370	400	0.90
Ivory Coast	70	2.5	180	20	60	260	300	0.90
Indonesia	120	1.7	210	60	120	390	500	0.80

Source: Country studies (Chapter 3-10)

Table 2.9 gives the annual production costs per hectare in the countries for the major costs elements. The variable costs can be divided into three components:
- labour and material inputs, such as fertilizers and fungicides
- fixed costs, which consist of the costs of depreciation and interest on equipment
- administrative costs and establishment costs in the form of an annuity (over 20 years).

The contribution of the three major cost categories varies between countries, but the distribution of costs between labour, materials and overhead costs is about 40, 25 and 35 % for arabica coffee and 60, 15 and 25 % for robusta coffee. For the selected countries, the three categories of costs per hectare are usually lower for robusta coffee than for arabica coffee. The cultivation of arabica generally requires some 150 to 250 man-days of labour, against a mere 100-120 man-days for robusta.

The coffee estates in Kenya are high-cost producers, for all cost elements. The labour inputs are higher than of other coffee producers, mainly through attention to pruning and spraying, but also partly through irrigation, processing and general maintenance. The high rate of fertilizers and sprays used is reflected in the high costs of material inputs; the overhead costs include many different elements, such as management fees and staff salaries, transport, interest payments and the depreciation and interest on production and processing equipment.

The production costs per hectare are also high in Colombia and Costa Rica, the two countries that have adopted intensive production with a high planting density.

Annual labour inputs are still moderate, but establishment costs and overhead costs are considerable. In Costa Rica, the rate of fertilizer is generally higher than in Colombia, largely explaining the higher yields and the difference in the costs per kilogram. But in Colombia, the first-stage processing is carried out on the farm and their costs are included in the production costs. With average yields only about 50 % of those in Costa Rica, the low-cost arabica producers in Brazil and Rwanda, have about the same costs per kilogram as in Costa Rica. Because of the extremely low yields resulting from intercropping and neglect, the costs of production of arabica coffee per kilogram in Cameroon are very high.

Despite the low average yields of robusta, the cost price per kilogram is about 25 % lower than that of low-cost arabica-producers, due to modest labour inputs and low material inputs and fixed costs. Since the prices on the world market for robusta coffee in the 1970s were seldom less than 75 % of the prices for unwashed arabicas and other milds, robusta-producers are generally at an advantage over arabica-producers.

Given the price levels on the world market (Section 2.5.1) and the production costs shown in Table 2.9, the best results are obtained by Costa Rica, Kenyan smallholders and the three selected robusta-producing countries. On the basis of the 1982/83 producer prices (Table 2.15) and the average costs and yields given in Table 2.9, the net return per man-day was by far the highest in Costa Rica (about US$6) and lowest in Indonesia and Cameroon (around US$2). In the other countries, the net return per man-day was about US$3.

These net returns per man-day are low; in several countries, they are just about or below the wages offered to day workers and certainly below the salaries and social charges accounted in the formal sector and on estates. Since (arabica) coffee is generally grown in densely populated zones with few alternative agricultural activities, these low net returns per man-day are accepted. For the same reasons, governments can impose considerable taxes on coffee, resulting in low producer prices. But in years of average and low prices on the world market, the net returns in several countries are insufficient for proper maintenance and reinvestment.

2.3.3 Alternative crops and other agricultural enterprises

The great importance of coffee for many exporting countries, results in heavy dependence on the world market. Many countries therefore feel a strong need for diversification. The national efforts were complemented by ICO, which, in view of the oversupply, has emphasized diversification, in particular in Central America and Colombia. Such programmes have had limited success, despite the many crops considered (e.g. cocoa, sugar-cane, food and oil crops, fruit and spice trees).

But in countries like Colombia and Costa Rica, few crops were found in the coffee zones that had the same ecological requirements, similar agro-economic characteristics and a world market of similar magnitude. In recent years, there is a more comprehensive approach to diversification in those areas, including promotion of various crops and of livestock, (agro-)forestry, agricultural processing and other off-farm activities.

Alternative crops for arabica growers are sometimes sugar-cane (Latin America) or oil crops, such as soya beans or sunflower, and occasionally tea in rather wet coffee zones of high altitude and cotton in dry zones at lower altitudes. In a country like Kenya, the tea and coffee zones overlap to some extent. Whereas tea offers more regular employment, the returns to farmers are usually less (Table 2.10). On farms with robusta coffee, such crops as cocoa (West Africa), cloves, pepper and coconuts (Indonesia) are found, and sometimes rubber and oil palm. Though not having exactly the same ecological requirements, there appeared to be competition between cocoa and coffee in Ivory Coast, according to the producer prices.

In Table 2.10, productivity on tea, coffee and cocoa plantations is compared on the basis of available data from smallholdings in Rwanda and Ivory Coast. Whereas the two upland crops show a higher land productivity, the lowland crops combine a lower land productivity with a better labour productivity. Similar figures are presented by Andreae (1980) in his comparison of the same crops in four different countries. He concludes that the two upland crops are highly suitable for overpopulated developing countries with low wage rates and the lowland crops for for more thinly settled areas with higher wages. That corresponds well with the actual situation in Rwanda and Ivory Coast.

In most countries, and in particular in the densely populated coffee zones of Rwanda, Kenya and Cameroon, coffee faces fierce competition from food crops. Since smallholders in many countries are inclined to interplant their coffee with food crops or another tree crop for various reasons, agricultural research organizations in recent years have paid much attention to intercropping. In Ivory Coast,

Table 2.10. Productivity in tea, coffee and cocoa plantations in two selected countries (1982).

	Tropical uplands (Rwanda)		Tropical lowlands (Ivory Coast)	
	tea	coffee arabica	coffee robusta	cocoa
Establishment costs (US$/ha)	.	1150	750	525
Labour requirements (man-days/ha per year)	450	300	80	70
Gross returns (US$/ha)	650	1200	300	400
Gross labour productivity (US$/man-day)	1.45	4.00	3.75	5.70
Agricultural wage rate (US$/man-day)	1.20	1.20	2.50	2.50

Source: Country studies (Chapter 7 and 9)

crops like upland rice, yams and groundnuts have proved suitable for intercropping, whereas bananas and maize were less suitable.

Legumes, including trees such as *Leucaena* have an advantage for intercropping in competing less for nitrogen and indirectly supplying nitrogen. Leguminous trees are also used for permanent shade; fruit and spice trees are generally less suitable. Black pepper vines growing up shade trees have the advantage of supplying additional produce. Bananas and plantains are often found in coffee plantations. They are sometimes used as a nurse crop (shade for young trees) and are subsequently left amidst the mature coffee trees.

Both smallholder and large-scale coffee producers (including estates) are often involved in some livestock activities, which compete with coffee for some inputs (labour at harvest time and grass for mulch) but which make use of marginal land and provide regular employment and income.

2.4 Processing, marketing and services

2.4.1 Processing and marketing channels

The organizational structure of the coffee sector is widely diverse in the producing countries, largely for historical reasons, differences in production structure, the share of coffee in the national economy and for political reasons. In most countries, production of coffee was originally undertaken by large farmers and business men, who either took care of export themselves or sold their produce to international dealers. Gradually marketing boards and producer associations entered the scene, and many of these were given wide powers in the marketing, pricing and export of coffee (Section 1.5.2).

The methods of processing and marketing depend on the commercial type of coffee produced (unwashed or washed arabica, or robusta), and on the size and type of production units. The producers of robusta coffee and unwashed arabicas usually take care of the processing themselves, since it involves only drying. In Brazil, coffee is thereafter sold to about 3000 private millers, who perform a central role in the marketing of coffee, whereas in Indonesia, the coffee is hulled on the farm or in the village (for a small fee, often in kind) and then sold through traders to a large number of exporters (250). In Ivory Coast and in Cameroon, cooperative organizations collect the dried robusta cherries. In both countries, they are pre-cooperatives, which merely bulk the produce of small coffee growers for an officially fixed small fee or bonus, paid by the agents of exporters. In Ivory Coast, hulling of robusta coffee has been centralized and undertaken in 16 factories since the late 1970s, whereas in Cameroon, with only a third of the production of Ivory Coast, there are more than 200 small factories for this operation. But in Cameroon too, there is a trend towards large centralized plants.

The situation is different for washed arabica coffee, the processing of which has to start the day after harvesting, and for which depulping installations and fermentation and washing facilities are required. In Colombia, farmers generally take care of this first processing stage themselves but they produce about 3 tonnes of green coffee equivalent per year. After this processing, they sell the parchment, partly through cooperatives to either agents of exporters or to centres of the national coffee organization (Federacafé). There are at least a hundred hulling factories.

For the estates in Kenya, the processing of cherry coffee into parchment is also an integral part of the production process, but the smallholders there, producing on average only about 150 kg coffee per year, were obliged to join a cooperative society to achieve economies of scale and to make sure of proper processing, by trained technicians. The drying is also undertaken by the cooperatives, which therefore require a considerable piece of land, next to a water point and need help from farmers and their families. The cooperative societies and unions received considerable assistance, in particular in cooperative and factory management. The estates and cooperatives deliver the parchment to the large central hulling factory of the Kenya Planters Cooperative Union in Nairobi.

In Rwanda, farmers produce only some 60 kg coffee per year and depulping is undertaken at small stations operated by groups of farmers themselves under supervision of the national coffee organization OCIR. Since these small centres lack large basins and hardly undertake any fermentation, the coffee is considered as semiwashed. The farmers dry the parchment by their homestead and sell it to traders, who take it to one of the three hulling plants.

In the arabica zones of Cameroon, the first-stage processing is also done at local centres but the coffee is also fermented there (fully washed). In Cameroon, the arabica growers also dry the parchment on the ground or on platforms next to their houses. Once dried, the parchment coffee is taken to the large cooperatives, which take care of the hulling and grading. The cooperatives in Western Province are also engaged in export of arabica coffee.

A complete different situation exists in Costa Rica, where all processing activities, from depulping, washing and drying to hulling, grading and bagging, are undertaken in about 100 private or cooperatively owned processing factories ('beneficios'). These sell the green coffee to about 25 exporters; some firms are involved both in processing and in export. The Costa Rica coffee industry faces the organizational problems of a rapid collection and transport of the (bulky) cherry coffee and the disposal after processing of large amounts of coffee pulp. There is the risk too of declining quality, if the cherries are not processed within 36 h of harvesting, but the set-up has also many positive features. Once the cherry coffee has reached the reception area of the factory, it passes all processing stages in one plant and is also prepared for export. Transport, bagging and physical losses are minimized, and quality control is facilitated.

Hulling factories in several other countries have a similar position. In Brazil, these mills are numerous and relatively small, whereas Ivory Coast now has 16 large factories, well distributed over the coffee areas. Kenya has one large plant in Nairobi.

The processing (hulling) factories do not always play a leading role in the marketing of coffee. There are also exporters, licensed by the governments. In all selected countries except Indonesia (or in all countries where coffee is a major export earner), a strong parastate organization is charged with regulating the marketing and export of coffee.

Because of large differences in the organization of processing, marketing, processing methods and processing installations, and because processing is undertaken by producers, traders, exporters or parastatals, it is difficult to compare the processing costs and efficiency. On the basis of information obtained in Kenya about pulping installations, and storage and grading facilities, and in Ivory Coast and Cameroon about hulling factories, rough estimates for investment in processing and grading facilities for a total production of 100 000 t green coffee (1982 prices) are as follows:

	Number	Investment (US$ million)
Two disc-pulping installations	1500	15 (washed coffee only)
Large hulling factories	4-8	14-24
Storage and grading complex	5-10	10-15

The transition in some countries from village level or small-scale hulling to large factories, employing some 50 permanent and 100 seasonal workers, has reduced employment in coffee processing, improving coffee quality and reducing losses. Generally the processing and marketing stages bring an additional employment of about 20 % over that in coffee production (including on-farm processing). In Brazil and Colombia, the ratio is lower and in Indonesia, with its long marketing channels, it is higher.

Information on processing and marketing costs and margins in each country is given in Tables 2.14 and 2.15 (Section 2.5.3). The costs of hulling, grading and sorting vary from US$0.02 per kilogram in Brazil and Colombia to US$0.14 per kilogram in Ivory Coast and Indonesia. The costly centralized hulling operations in Ivory Coast have greatly reduced the costs of handling and administration in the village, which are still considerable in Indonesia.

Though Colombia, Ivory Coast and Indonesia have one or more soluble coffee plants (several of which partly or fully owned by Nestlé), Brazil is the only coffee-producing country with a major soluble industry. In 1975, when 8 of the present 11 factories were in operation, the total capital invested in these plants was about US$42 million and those then employed 2800 people (for the processing of about 120 000 t of green coffee equivalent). In the period 1978 to 1983, exports of soluble coffee constituted about 15 % of Brazilian coffee exports.

2.4.2 *Markets and marketing functions*

There is a considerable domestic market for coffee in the 3 Latin American countries, where consumption of coffee per person is reasonably high (about 4 kg per year) and in Indonesia, because of its large population. In those countries, there are many roasting plants. In the African countries, more than 95 % of the coffee production is exported and the roasting and retail marketing of coffee are negligible. Since coffee can generally be better stored and transported in the form of dried parchment or green coffee than roasted, and importers prefer to mix coffee from different origin, hardly any coffee is exported after roasting (and grinding).

In most of the selected countries, about 20-60 firms export coffee. On average, they export about 3000-15 000 tonnes per year. In Indonesia, there are at least 250 registered exporters, but only 25 of them export large amounts. In Rwanda, only one firm is involved in export of about 25 000 tonnes coffee per year. In Kenya, coffee is sold to exporters through public auctions held weekly. In Costa Rica, auctions play a role in the sales of coffee allocated for the domestic market.

Coffee exports are generally well distributed over several importing countries, but more than 50 % of Rwanda coffee is sold to the United States and almost 50 % of the Rwandan coffee is exported to West Germany. The United States is also the major market for Brazilian and Indonesian coffee, and West Germany for coffee from Colombia and Costa Rica. France is still the major importer of coffee from Ivory Coast and Cameroon.

The main marketing functions in the coffee trade are collection, transport, storage, grading and bagging. Pricing and financing are related activities, in which Governments and parastatals play a major role (Section 2.5). Wholesale and retail activities for domestic sales have only been covered in this study to a limited extent. With the gradual decline in the coffee estate sector and the emergence of smallholder coffee farmers, appropriate collection systems have become important. The costs of collection are generally high in smallholder production areas, whether use is made of cooperatives (Kenya, Cameroon) or of private traders (Rwanda). In Tables 2.14 and 2.15 in Section 2.5, these costs fall partly under administrative costs (cooperatives) and partly under transport and handling costs.

At the end of the 18th Century, the coffee plantations expanded as the railways were built in many countries, which facilitated the establishment of arabica in the appropriate ecological zones. But having a favourable value-weight ratio, coffee is now mainly transported by road. During the war in Uganda, Rwandan coffee was transported by air to Mombasa in Kenya, the nearest port. The Brazilian and Colombian coffee organizations have their own shipping fleet.

Storage or warehousing has always played a major role in the coffee trade, in particular in Brazil, which dominated the trade and kept stocks to influence the world prices. Under the International Coffee Agreements, warehousing and the verification of stocks have become important elements in organizing the international coffee market and in stabilizing prices on the world market. The warehousing capacity is enormous in Brazil, which can keep stocks of 7 million tonnes of coffee, more than world production in one year. Colombia and Ivory Coast also have a considerable storage capacity, given their average annual production. A large proportion of total stocks are held by the parastate coffee institutes or their subsidiaries. Since those organizations are involved in many other activities, it is difficult to estimate the considerable costs of this warehousing.

Although there are some international classification systems in coffee trade for type (number of defects) and description (size, taste), there are still large differences in the grades and quality classes distinguished in the various production countries. Most countries have 5 to 6 grades, according to the size of the beans that pass certain screen holes: from 4 mm (screen 10) to 8 mm (screen 20) diameter. But the number of quality classes ranges from 4 in Colombia and Cameroon to 10 in Kenya. The number of defects and the cup quality are major criteria. In Costa Rica, coffee is classified on the basis of the ecological conditions under which it is grown.

2.4.3 Supporting services

In most countries where coffee is a major export commodity, coffee growers benefit from special extension services, credit and supply of inputs. The national coffee institutes usually play a major role in provision of these services or have delegated them to local organizations (Colombia).

In Brazil and Colombia, provision of credit is one of the major supporting services offered to farmers by the coffee institutes. In Costa Rica, farmers obtain credit mainly from the processing factories, through which they also receive their first and deferred payments for coffee deliveries. In Kenya, small coffee growers can obtain credit through the cooperative system. In the other African countries, farmers can qualify for certain subsidies on planting stock, spraying materials and fertilizers.

Extension services are the responsibility of either the Ministeries of Agriculture or the parastatals, or both (e.g. Rwanda). The parastate organization in charge of the technical assistance to cocoa and coffee farmers in Ivory Coast, SATMACI, has gradually become a regional development organization for the whole cocoa and coffee zone, covering about a third of the country. In Cameroon, large coffee cooperatives such as Uccao and Zapi-est now make a major contribution to general extension services in their regions. In Kenya, extension services are the sole responsibility of the Ministry of Agriculture, but there are coffee officers (subject specialists) in the coffee zones, who have close links with the coffee research station. In Indonesia, the Directorate of Estate Crops is responsible for extension services and, in 1979, packages for the promotion of export crops were established.

In view of developments and reorganizations in several countries in the 1970s, most countries are still seeking the best solutions to provide supporting services to small farmers.

Both in its organization and in its orientation, research on coffee varies considerably for the eight production countries. But in all these countries, there is at least one organization involved in coffee research. In Brazil, the three major coffee states each have their own research institute, but the one in Campinas is the best known, in particular for its breeding work. The Chinchina Coffee Research Station in Colombia is sponsored and administered by the Federation of Coffee Growers, and is involved in a wide range of research activities. The emphasis there has been on more intensive cultivation systems with compact varieties (e.g. 'Caturra'). In Costa Rica and other Central American countries, a cooperative programme for the improvement of coffee production has been established (Promecafé), which has focused research on varieties and hybrids resistant to leaf-rust.

In both Colombia and Costa Rica (Central America), there has been much reseach also on suitable diversification crops. In Cameroon and Ivory Coast, research on coffee is done by sister organizations of the French Coffee and Cocoa Institute (IFCC), which has been particularly involved in the vegetative selection of robusta coffee, and in Ivory Coast, in the development of arabusta coffee. In Indonesia, one of the agricultural research stations, formerly linked to the estate sector, has become well known for its coffee research. Rwanda has no separate coffee research station, all agricultural research being done by one organization (ISAR). In Kenya, one organization is exclusively engaged in the research on coffee: the Coffee Research Foundation, financed by the Coffee Board. It is particularly known for research on coffee berry disease. In recent years, it has been involved in socio-economic surveys among smallholders and estates.

2.5 Export earnings, pricing and taxation

2.5.1 Position on the world market

The demand by importing countries for a particular type of coffee depends on their blending and roasting methods, on the type of brew preferred, and on consumption per person. The United States is still the major single market, in particular for medium-quality arabicas, despite rapidly declining consumption per person there in the 1970s through competition from soft drinks, juices and tea (Table 2.11). The Scandinavian countries, West Germany and the Netherlands are large markets for the higher-quality coffees: Colombian milds and other milds. The value of their imports is relatively high (Table 2.11).

Because of its rather flat taste and its high content of caffein, 100 % robusta coffee is seldom used, though most countries use a fair amount of it in their blends. In particular, France, Spain and Japan import large amounts of robusta. In the United States and the United Kingdom, much instant coffee is used, which

Table 2.11. Proportional distribution of world imports of coffee by ICO members and comparative level of unit value of imports (world average = 100).

	World imports (%)			Share of robusta* (%)	Level of unit value of imports (1975/76 to 1979/80)
	1971/72	1975/76	1979/80		
United States	43	36	33	24	96
EEC	37	40	42	37	101
West Germany	11	11	13	.	107
France	8	9	9	.	99
Italy	5	6	6	.	95
Netherlands	4	5	4	.	104
United Kingdom	4	4	4	.	89
Sweden, Finland & Norway	6	6	6	5	108
Japan	2	4	5	37	106
Spain	2	3	3	34	106
Canada	3	3	3	18	104
Other	7	8	8	23	96
Total/average (%)	100	100	100	29	100
(million t)	3275	3619	3648	.	.
Green coffee	.	92	91	.	100
Roasted coffee	.	2	2	.	115
Soluble coffee	.	6	7	.	94

Source: ICO (1977-1983)

is predominantly made from robusta. Low-quality coffee is generally disposed of on the domestic market and to some extent on the non-quota markets, representing countries that are not members of the International Coffee Organization. A large expansion of existing markets is impossible and new markets can only be found in Asia (China).

Major net importers of processed coffee (mainly soluble) are the United States, the United Kingdom, Canada, Japan and France. Although several producing countries have one or more soluble plants, Brazil is the only coffee-producing country with a large production of instant coffee. In contrast to the product made in Europe and the United States, the Brazilian instant coffee consists largely of arabica coffee. West Germany, Switzerland and the Netherlands have large coffee processing industries and export considerable amounts of processed coffee. Another factor of importance in the trade is the season of the year when coffee is sold on the world market. In most producing countries, the crop year starts in October. But in Brazil, Indonesia, Ecuador, Peru, Madagascar, Rwanda and Burundi, it starts in April. These cropping years influence prices to some extent. Coffee can be and generally is stored for some time, but climatic conditions in producing countries and particularly in harbour towns in the (lowland) tropics affect the

Table 2.12. Coffee exports by selected countries (ICO members) to all destinations (1000 t) and the proportional distribution.

	Crop year (October-September)				Proportional distribution
	1972/73	1975/76	1978/79	1981/82	(%)
Unwashed arabicas	1191	863	884	1106	27
Brazil (some robusta)	1099	781	793	1008	25
Colombian milds	513	561	808	700	18
Colombia	375	421	686	539	14
Kenya	72	81	74	102	2
Other milds	919	923	1121	969	27
Costa Rica	87	61	94	93	2
Rwanda	15	32	21	26	1
Robustas	980	1065	990	1030	27
Cameroon	73	110	94	95	3
Indonesia (some arabica)	93	130	234	212	5
Ivory Coast	217	332	277	288	7
Total	3603	3412	3802	3805	100

Source: ICO (1977-1983)

quality. Exporters are not always in a position to hold stocks for long periods and want to sell their quota in time. By the apportionment of the annual quota imposed by the ICA into quarterly quotas, the seasonal influence is mitigated but it still plays a certain role, in particular in years when there is a shortfall in supply, after a frost in Brazil or when there is a bumper crop there.

The coffee-producing countries compete with each other on the world market, in particular when they sell the same type of coffee. Table 2.12 lists the world's coffee exports by type for each of the countries over four years during the period 1972 to 1982. Over this whole period, total exports of unwashed arabicas, other milds and robustas were about equal, each with 27 % of world exports. Colombian milds constituted only 18 % of total exports. Brazil and Colombia are the predominant producers within their groups, whereas many countries produce other milds and robustas.

Table 2.13 shows how far different types and qualities of coffee fetch different prices on the world market. The highest value of exports was during the period 1972 to 1982 by producers of Colombian mild coffee: about 20 % higher than the world average. Within that group, Kenyan coffee invariably fetched higher prices than Colombian coffee, probably because of the organization and methods of processing and grading. Rwandan coffee is in fact semiwashed (washed with little fermentation) and is therefore considered slightly inferior to 'other milds'. The lowest average value during the period 1972 to 1982 was obtained by the group of robusta-producers. However after the 1975 frost in Brazil, when prices rose sharply because of the general scarcity, demand switched towards the cheaper

Table 2.13. Export unit value in selected countries (ICO members) to all destinations.

	Crop year (October-September)				Average index
	1972/73	1975/76	1978/79	1981/82	
Unwashed arabicas	1.11	1.97	2.90	2.00	94
Brazil	1.30	2.58	3.49	2.78	127
Colombian milds	1.35	2.23	3.32	2.86	119
Colombia	1.41	2.17	3.34	2.91	120
Kenya	1.30	2.58	3.49	2.78	127
Other milds	1.14	2.07	3.03	2.61	107
Costa Rica	1.15	1.99	3.01	2.63	109
Rwanda	0.70	1.57	3.04	2.78	91
Robustas	0.93	1.57	2.87	1.86	87
Cameroon (also arabica)	1.10	1.45	2.95	2.09	89
Indonesia	0.78	1.48	2.70	1.50	86
Ivory Coast	1.03	1.75	3.12	1.99	98
Average of total	1.10	1.91	3.02	2.28	100 (2.39)

Source: ICO (1977-1983)

types and robusta coffee fetched about the same price as unwashed arabicas.

The Associate Countries of the European Community in Africa are at an advantage, having duty-free access to the European community, whereas other countries have to pay the common external tariff of 5 % ad valorem. The Associates (ACP) in Africa, the Caribbean and the Pacific also benefit from the Stabex scheme, through which sharp falls in income from primary commodities, such as coffee, are compensated.

2.5.2 *Pricing and taxation*

World trade in coffee is a typical south-north trade. In the period 1979-1981, 95 % of the coffee on the world market was exported by less developed countries and 94 % was imported by developed countries (Herrmann, 1983). In that period, export earnings from coffee by the developing countries ranked highest of all their agricultural export commodities, followed by sugar, rubber and cocoa products. For most of the 60-odd developing coffee-exporting countries, coffee is the major export earner.

So the governments of most of these countries try, on the one hand, to support the producers, in order to maintain or increase production levels, and, on the other hand, impose taxes and regulate exports, in order to utilize export and foreign exchange earnings to the full. To what extent both types of intervention are successful depends largely on prices on the world market. At the annual fixation of minimum producer prices, subsidies and export taxes, the likely level of world prices has to be taken into account. Unfortunately world trade in commodities such as coffee, tea and cocoa is highly unstable.

Herrmann (1983) showed that the price instability of these three commodities ranked above, and the quantity instability below the mean value for some other agricultural export commodities in the period 1969-1981. The instability of export earnings ranked highest on the markets of coffee and cocoa.

To reduce the risk of fluctuating prices for the producer, coffee and cocoa exports in some West African countries (e.g. Ivory Coast) are controlled by a so-called 'Caisse de Stabilisation', a parastate organization that aims at internal price stabilization. As a result, the instability of producer prices for coffee in those countries in the period 1969-1981 was considerably less than the instability of export prices. Producer prices were kept low, and in most years, particularly during periods of high prices on the world market, large funds accrued to the parastatal, which also controls marketing and export margins. Some of these funds are used for price stabilization of other major commodities, for investment in rural institutions and for the rural infrastructure. The coffee farmers only benefit indirectly and to a limited extent from these investments.

The Brazilian Coffee Institute and the Federation of Coffee Growers in Colombia both operate alongside private organizations in the marketing and export of coffee, through a system of minimum guaranteed producer prices, and detailed regulations for export of coffee by private exporters. These exporters have to submit a considerable part of their export earnings, mostly in foreign exchange to the Government, through these parastate institutions. Much of the funds thus created are ploughed back into the coffee sector, through subsidized credit to coffee farmers and a multitude of investment projects in the coffee zones.

In most of the selected countries, export taxes ad valorem are imposed, the rate of which depends in some countries on prices on the world market. In Costa Rica, this tax ranges by law between 5-18 % and alteration in it helps to stabilize domestic prices. Export duty in Kenya varies too with prices on the world market but hardly ever exceeds 10 % and so has little stabilizing effect. In Rwanda, a progressive tax on coffee exports was abolished in 1979 and a formula has been applied, in which the export earnings per unit above a certain level accrue partly to stabilization fund and partly to the treasury. At low prices on the world market, the Government then subsidizes the coffee.

Apart from contributions to stabilization funds and export taxes, all countries impose small taxes on the various participants in production, trade and domestic consumption. Many of these (minor) turnover and sales taxes are found in Costa Rica and Indonesia.

2.5.3 *Distribution of income from coffee*

On the basis of export prices (Section 2.5.1), domestic prices, margins and taxes, and various cost elements in production, processing and marketing, we can

Table 2.14. Proportional distribution (%) of export unit value by cost elements and by recipient in the selected countries (1982).

Category	Bra.	Col.	Cos. R.	Kenya		Rwa.	Cameroon		Ivo. C.	Ind.
				smallh.	est.		arab.	rob.		
Port handling, exporters	2	2	2	3	3	3	2	2	2	8
Transport, storage, handling	2	3	5	9	4	8	5	7	5	21
(Wet) processing[1, 2]	–	–	3	8	–	–	–	–	–	–
Hulling, grading, sorting	1	1	4	2	2	2	5	4	6	8
Administration and finance	1	1	2	6	1	1	6	5	2	9
Taxes[3, 4]	47	41	22	7	7	18	11	10	10	5
Stabilizing & other costs			–	–	–	5	26	25	27	–
Producer payments	47	51	62	65	83	63	45	47	48	49
Total	100	100	100	100	100	100	100	100	100	100
Producers	47	51	62	65	83	63	45	47	48	49
Cooperatives	–	1	–	18	–	–	14	8	–	–
Traders, exporters, industries	6	6	15	8	8	14	4	10	15	46
Governmental funds[3, 4]	47	42	23	8	8	23	37	35	37	5

1. Processing undertaken by producers themselves (Brazil, Colombia and estates in Kenya).
2. Processing undertaken at local depulping stations (Rwanda and arabica in Cameroon).
3. Including contribution quota in Brazil.
4. Including reintegro in Colombia.

Source: Country studies (Chapter 3-10)

compare the share of export earnings received by the different participants in the coffee sector.

Table 2.14 shows the relative distribution of the export value by cost elements and by recipient. Marketing costs and margins are relatively low in Brazil and Colombia, where large amounts are handled. These costs and margins are high in Indonesia and in smallholder areas of Kenya, Rwanda and Cameroon. Taxes and transfers to stabilization and other funds constitute a major cost element in the marketing and export of coffee in most countries, except Kenya and Indonesia. In Costa Rica and Rwanda the taxes as such are quite high but other transfers are limited.

Through the various deductions for marketing costs and taxes, the share of producers in export earnings is rather low in Brazil, Colombia, Cameroon and Ivory Coast. This is also indicated by the nominal protection coefficient (Table 2.15), which is defined as the ratio between producer price and border price adjusted to the farm gate. Since no allowance is made in this coefficient for exchange rate deviations (not excessive in these countries), the figures are only indicative for the degree of protection offered by government intervention. The recycling of funds in the coffee sector of Brazil and Colombia, and to a much lesser extent in Ivory Coast and Cameroon, has not been taken into account. That would raise the nom-

Table 2.15. Prices, marketing costs, intervention and efficiency factors (US$ per kg) in the selected countries (1982).

		Bra.	Col.	Cos. R.	Kenya		Rwa.	Camoon		Ivo. C.	Ind.
					smallh.	est.		ara.	rob.		
a.	Export price (f.o.b.)	2.55	2.90	2.64	2.94	2.90	2.72	2.47	2.15	1.90	1.75
	Export margin	0.05	0.02	0.04	0.07	0.07	0.03	0.01	0.01	0.01	0.08
	Custom, harbour	0.01	0.02	0.02	0.04	0.04	0.01	0.04	0.04	0.01	0.06
	Administration & financial costs	0.02	0.02	0.04	0.16[1]	0.02	0.01	0.17[1]	0.11[1]	0.03	0.15
	Processing, grading & sorting	0.02	0.02	0.10[2]	0.20[2]	0.04	0.05	0.12	0.09	0.14	0.14
	Transport, storage & handling	0.05	0.10	0.20	0.35	0.09	0.26	0.13	0.14	0.10	0.37
	Total	0.15	0.18	0.40	0.82	0.26	0.36	0.47	0.39	0.29	0.80
b.	Border price equivalent	2.40	2.72	2.24	2.12	2.64	2.36	2.00	1.76	1.61	0.95
c.	Actual producer price	1.20	1.52	1.64	1.90	2.42	1.73	1.11	1.00	0.91	0.86
c/b	Nominal protection coefficient	0.50	0.56	0.73	0.90	0.92	0.73	0.55	0.57	0.56	0.91
d.	Cost of domestic resources	0.87	1.49	1.05	1.43	1.19	1.11	1.40	0.96	0.93	1.22
e.	Traded inputs	0.48	0.37	0.44	0.65	1.14	0.51[3]	0.32	0.19	0.12	0.21
a-e	Net foreign exchange earnings	2.07	2.53	2.22	2.29	1.76	2.21	2.15	1.96	1.78	1.54
d/a-e	Domestic resource cost coefficient	0.42	0.59	0.47	0.62	0.68	0.50	0.65	0.49	0.52	0.79

1. Including cooperative administration.
2. Including wet processing.
3. Including foreign exchange costs of trade to Mombasa.
Source: Country studies (Chapter 3-10)

inal protection coefficients.

The modest intervention by the Indonesian and Kenyan Governments is also reflected in their 1970-1979 supply response to changes in world price (0.44 and 0.28, respectively). The other countries showed elasticities of 0.15 or less (Josling, 1984).

Table 2.15 also gives the domestic resource cost coefficient, defined as the ratio of the cost of domestic resources used in producing coffee to the net foreign exchange earned. This indicator shows that all the selected countries are at a comparative advantage to produce coffee by existing technologies. But there are differences between the countries, with the lowest coefficient (most advantageous position) for Brazil and Costa Rica, and the highest for Indonesia. For absolute measures of the comparative advantage of producing coffee, one must calculate the coefficients for other commodities in the countries.

In the 1970s, there were enormous fluctuations in the terms of agricultural trade for developing countries, through the instability of agricultural commodity prices and sharp increase in petroleum prices, that also greatly influenced the prices of manufactures and agricultural inputs (e.g. fertilizers). These changes also affected the 8 selected countries, though in the period 1970-1979 this group fared relatively well (Josling, 1984). The average annual growth rates of the purchasing power of agricultural exports were about 2 % for Kenya and Brazil; 3 % for Costa Rica; more than 5 % for Cameroon and Indonesia; over 7 % for Colombia; nearly 10 % for Ivory Coast and over 15 % for Rwanda. These growth rates were definitively related to the coffee boom (1976-1978), which could also be held responsible for the high instability index in purchasing power of agricultural exports, for some of the selected countries: nearly 20 for Rwanda and Kenya and 17.5 for Brazil. Indonesia, where agricultural exports do not constitute a major part of total exports, and Cameroon, with well diversified exports and strong policies for stabilization of internal prices, showed a low instability index (below 10).

Though the International Coffee Organization did not halt the rapid increase in prices in the coffee boom, it has prevented prices from dropping excessively in the period 1980-1982, which would have been disastrous for many producing countries relying heavily on their exports of coffee. But since the coffee boom stimulated coffee production in many countries, the stocks increased rapidly in the early 1980s, putting pressure on the international (ICA) prices and leading to 'special deal' arrangements and uncontrolled exports.

2.6 Conclusions

Coffee is a tropical product almost exclusively produced in developing countries, that does not (yet) face serious competition from direct substitutes produced in developed major consumer countries. Since its establishment as a major beverage, it has been considered an excellent export crop, that could raise tax revenues and earnings of foreign exchange.

In this publication, numerous factors have been distinguished that determine the extent to which countries, producers, traders and governments benefit from the production of coffee.

The world's coffee market distinguishes between different types and qualities of coffee, resulting in such 1982 export price differentials of US$2.90 per kilogram for high-quality arabica coffee against US$1.80 for low-quality robusta coffee. The choice between arabica and robusta coffee largely depends on ecological factors and to some extent on the occurrence of some major diseases. Arabica production is therefore largely confined to the usually densely populated tropical highland zones. There are more opportunities for expansion in robusta coffee at low and medium altitudes, where it faces some competition from other tropical export crops (e.g. cocoa).

More than 90 % of the coffee is exported as green coffee. It is blended and roasted in the consumer countries. Only Brazil exports a considerable amount of soluble coffee. Roasting and processing into soluble coffee generally require a capital-intensive and not so labour-intensive technology.

Although such material inputs as fertilizers and chemicals for disease control are important to increase productivity, land and labour still constitute the main production factors in most countries. Depending on their relative scarcity, these factors are combined in different ways: Brazil uses 1.7 ha and 0.5 man-equivalent for the production of 1 t of green coffee, whereas Rwanda uses 1.4 ha and 1.9 man-equivalent. The planting stock is also important. With planting density and the fertilization, it has a major influence on yield, of special importance in densely populated arabica zones.

The area under coffee per farm ranges in the case studies from as little as 0.1 ha for the farmers in Rwanda to over 50 ha for the estate sector in Kenya and more than 10 ha for the plantations in Brazil. The number and importance of large coffee estates has generally decreased and coffee has become a typical smallholder crop in many countries. Because of their high overhead costs, estates are often high-cost producers. Since harvesting, which requires half the labour input, takes about 3 months or less, other activities need to be undertaken during the remainder of the year, or seasonal labourers have to be relied upon. Small farms are usually considerably diversified.

The share of labour costs in total production costs is with present methods about 40 % for arabica (150-250 man-days/ha) and 60 % for robusta coffee (100-120 man-days/ha).

With average yields between 600 and 1000 kg/ha for arabica and only 400 kg/ha for robusta, costs per kilogram were between US$1.20-1.70/kg for arabica and around US$0.90/kg for robusta (or on average about 50 % of the export values) in 1982. The margin between production costs and export value is made up of the processing and marketing costs and of various taxes or transfers to price stabilization and other national funds. Under the prevailing (1982/83) producer prices, the net returns were in most countries only just about the level of wages offered to day workers. Since (arabica) coffee is generally grown in densely populated zones, with few alternative agricultural activities, these low net returns per man-day are accepted. The same factors explain why governments can impose considerable taxes, resulting in low producer prices.

The organizational structure of coffee processing and marketing varies markedly between countries (Section 2.4). Dry-processed robusta coffee only requires investment in hulling facilities and in grading and sorting, and the collection and marketing is usually undertaken by traders or agents of exporters. Medium-scale hulling factories in major production zones contribute to a better quality control. High-quality arabica coffee requires pulping facilities and washing basins near the producers. In smallholder areas, this has caused the formation of cooperative organizations responsible for both processing and marketing.

In many coffee-producing countries, parastate coffee organizations are involved in fixing guaranteed producer prices, and directly or indirectly also in fixing marketing margins. In almost all producing countries, there is a system of ad valorem or progressively increasing export taxes. These regulations have a certain stabilizing effect on internal prices. In years of high prices on the world market, they also lead to a large accumulation of funds by the national organizations, usually meant for price stabilization and for investment in the rural infrastructure. However allocation is not always based on economic criteria and often has inflationary effects.

Since the 1976-1978 coffee boom, ICO has stabilized prices on the world market of coffee, though unsold stocks accumulated in many producing countries. Diversification programmes in those countries have not generally shown singificant results.

References

Andreae, B., 1980. The economics of tropical agriculture. Commonwealth Agricultural Bureaux, Slough.

FAO, 1982. Production Yearbook. FAO, Rome.

Herrmann, R., 1984. Trade and trade policy on the world markets in coffee, cocoa and tea. Quarterly Journal of International Agriculture 23 (2): 129-148.

ICO, 1977-1983. Quarterly Statistical Bulletin on Coffee (several issues). ICO, London.

Josling, T., 1984. Agricultural prices and export earnings: The experience of developing countries in the 1970s. FAO Economic and Social Development Paper 43. FAO, Rome.

Landberg, H.E., editor in chief, 1976. World survey on climatology, Volumes 8, 10 & 12. Elsevier, Amsterdam.

World Bank, 1984. World development report 1984. Oxford University Press, Washington.

Part II. Country studies

3 Coffee in Brazil

3.1 Background

Brazil is by far the largest country in Latin America. Its area of 8 500 000 km^2 is greater than that of Europe or Australia. The total population in 1982 was about 127 million inhabitants. There are some ten cities with more than a million inhabitants, the largest being São Paulo and Rio de Janeiro with about 10 and 8 million, respectively. The capital Brasilia, newly built since 1956 already has about a million inhabitants. Administratively Brazil is divided into 22 states, the federal district Brasilia, and four federal territories (three in the Amazon area and the fourth consisting of the Fernando de Noronha Islands).

In 1982, the GNP per person was about US$2240. The share in the major sectors in the GDP is respectively: agriculture 13 %, industry 34 % and services 53 %. And of the total work force in 1982 of about 55 million, 30 % was working in the agricultural sector and 24 % and 46 % in industry and services, respectively (World Bank, 1984). Though since the 18th Century, the Brazilian shelf has produced several minerals, including gold, diamonds and now a considerable amount of iron ore, the mining sector is unimportant and petroleum products are largely imported. The world's largest hydro-electricity station has been recently built at Itaipu on the Parana River.

Table 3.1 shows the major crops in Brazil. In the period 1970 to 1980, the coffee sector contributed an average of about 10 % to agricultural production or about 1 % of the GDP. And 3 to 4 % of the total work force was engaged in the production of coffee. Until recently, coffee was the major export but since 1981 it has been passed by soya beans with 12 % of total exports, against 10 % for coffee (IMF statistics, 1983). Other important agricultural exports are sugar, oranges, cocoa beans and tobacco. It is estimated that at least a third of the area of the country is suitable for agriculture, but now only 7 % of the land area is used for arable and permanent crops, and 19 % is used as permanent pasture (FAO, 1981).

Since 1810, Brazil has been the largest producer of coffee in the world and until the 1940s accounted for more than half of world production. With an annual production of about 1.4 million tonnes, it still contributed about 27 % of total world production in the years 1979 to 1981. Its production consists predominantly of unwashed arabica but production of robusta coffee is increasing, and in 1982 constituted about 7 % of total production.

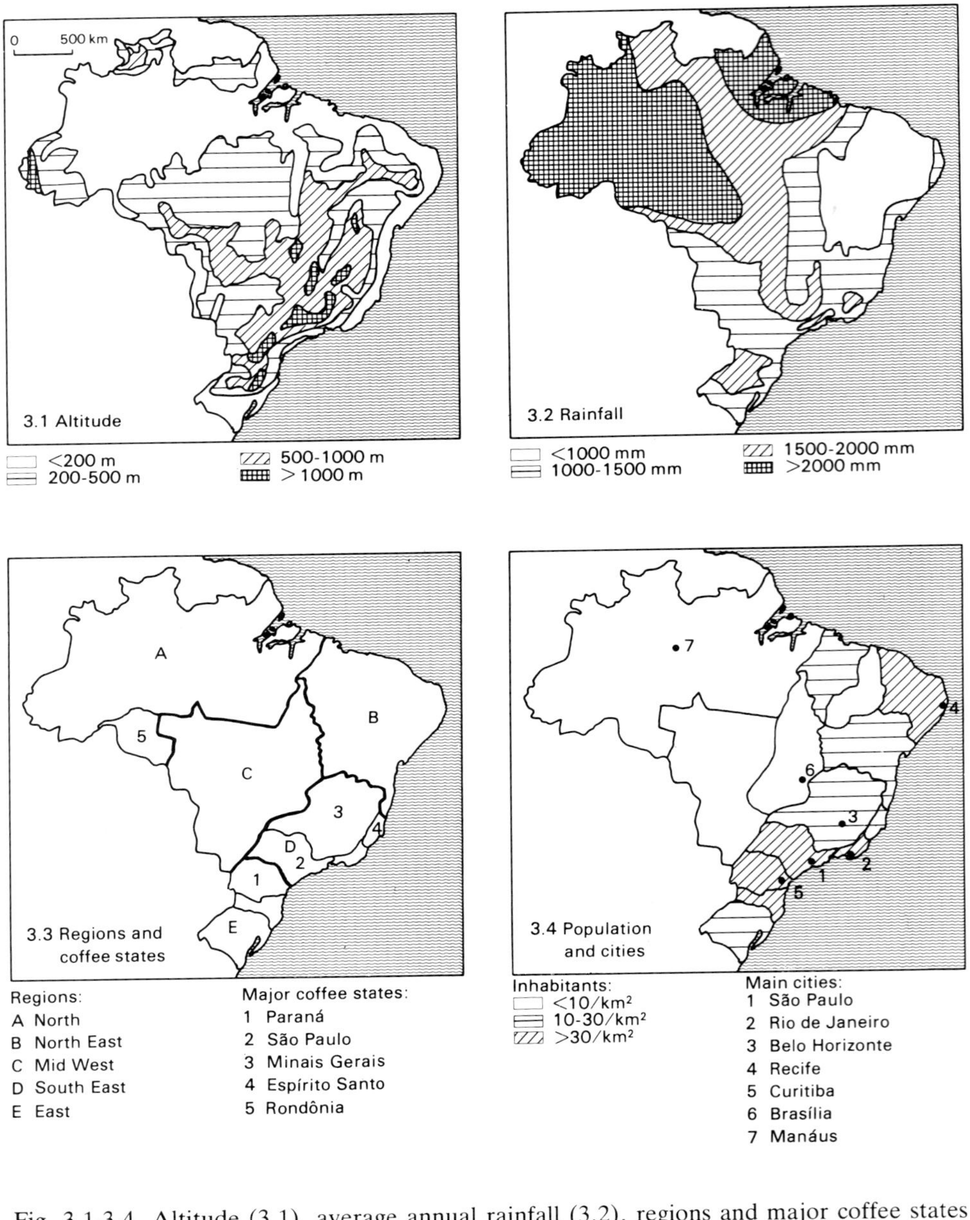

Fig. 3.1-3.4. Altitude (3.1), average annual rainfall (3.2), regions and major coffee states (3.3) and population density and major cities (3.4) of Brazil. After Hohé (1979).

Table 3.1. Average area harvested, yield and national production of major crops over 3-year periods.

Crops	Area (1000 ha)		Yield (kg/ha)		Production (1000 t)	
	1969-1971	1979-1981	1969-1971	1979-1981	1969-1971	1979-1981
Food crops						
maize	10021	11416	1365	1686	13680	19379
cassava	2042	2070	15655	11821	29922	24474
rice	4788	5909	1430	1442	6847	8535
beans (dry)	3685	4627	642	470	2366	2165
Oil crops						
soya beans	1314	8836	1178	1519	1547	13457
groundnuts (in shell)	670	281	1307	1539	876	433
coconuts	.	.	.	.	331	254
Beverages						
coffee	2455	2317	487	603	1197	1403
cocoa	443	475	412	667	183	317
Other crops						
cotton	2644	2050	691	880	1828	1805
sugar-cane[1]	1708	2648	3149	3106	5378	8225
oranges	180	450	14000	18560	2514	8352

1. Production of sugar centrifugal and non-centrifugal.
Source: FAO (1971-1983)

3.2 Ecology

The immense land area of Brazil consists mainly of large plains at an altitude of several hundred metres, alternating with mountain ranges and lowland. Almost 80 % of the land is below 500 m and only 4 % above 1000 m (Figure 3 1). The Brazilian Institute for Geography and Statistics distinguishes five main regions: the North, the North-East, the South-East, the Mid-West and the South (Figure 3.3). The Northern Region coincides largely with the Amazon watershed and is tropical rain forest (Selva), with average annual rainfall above 2000 mm. It contains lower (varzeas) and higher areas (terras firme). After deforestation, the soils of the higher areas may quickly lose their fertility by leaching from the thin layer of top soil.

The North-East Region is warmer and drier, and the rains are highly unpredictable. Part of it consists of the 'drought-polygon' where the vegetation is mainly composed of cacti ('caatinga'). The average annual rainfall is below 1000 mm. The inland zone of the South-East Region is characterized by a savanna climate (campos) with rains from September to May, and a total annual rainfall of 1000-1500 mm (Table 3.2 and Figure 3.2). This is the only mountainous region with peaks above 3000 m; its valleys have fertile agricultural land. The coastal areas of

Table 3.2. Distribution of rainfall for several coffee zones.

State	Station[1]	Altitude (m)	Annual rainfall (mm)	Monthly rainfall (mm)					
				J	F	M	A	M	J
Mato G.	Cuiabá	171	1378	213	200	222	106	46	14
Minas G.	B. Horiz.	915	1561	319	202	157	79	20	9
Espír. S.	Vitória	31	1410	151	118	150	130	90	66
Rio de J.	Rio de J.	90	1093	157	125	134	102	63	56
São Paulo	Campinas	663	1393	241	199	148	61	56	53
Paraná	Curitiba	949	1363	183	149	105	76	88	104
				J	A	S	O	N	D
	Cuiabá			9	27	48	124	161	208
	B. Horiz.			8	18	42	136	225	346
	Vitória			72	48	85	135	169	191
	Rio de J.			51	40	63	80	92	130
	Campinas			29	36	75	121	160	215
	Curitiba			69	85	124	122	120	138

1. Annual daily mean and extreme minimum temperatures (°C) for these stations are respectively: (25.5, 1.2); (20.6, 2.4); (23.2, 9.3); (22.7, 10.2); (19.7, −1.5); (16.4, −6.3).
Source: Schwerdtfeger (1976)

both the South-East and North-East Regions are tropical forest, where sugar-cane and cocoa plantations have been established since colonial times.

The Mid-West Region is a transitional zone, in the north resembling the Amazon Area and in the east the South-East savanna. Its features are a tropical continental climate, with a distinct dry period (May to August), some heavy showers from October to December and a long wet period from February to May. The savanna zone is divided in a zone with trees (campo cerrado) and without trees (campo limpio). This region also includes a large swampy area, the Pantanal.

The Southern Region largely has a subtropical climate. It contained large forests of *Araucaria* (a conifer) but some of these have already been eliminated. In some areas, night-frost occurs occasionally. The coastal strip contains a maritime tropical climate and the most southern lowland zone consists mainly of grassland (pampa).

Even although Brazil has always been the leading world producer of coffee, the climatic conditions in its coffee-growing areas are certainly not excellent. Areas where the climate is suitable for coffee are also suitable for some other crops. Arabica coffee thrives best at an average annual temperature of 18-23 °C without major fluctuations, which is found at an altitude of 1000-2000 m in most production countries (within the tropics). In Brazil, such areas are found only in Minas Gerais State (Figure 3.3), and production areas with similar temperatures are

Table 3.3. Registered grave frosts in coffee regions (1900-1981). Data from 1921-1928 are missing.

Date		Intensity	Interval in years
19 August	1902	very grave	20[1]
12 August	1904	grave	6
9 September	1912	grave	8
25 June	1918	very grave	16
29 June	1942	grave	21
5 July	1953	grave	11
2 August	1955	grave	2
21 July	1957	grave	2
6 August	1966	grave	9
18 July	1975	very grave	57
18 July	1981	grave	6

1. At least 20 years; registration dates back from 1882.
Source: Marshall (1983)

found at lower altitudes near to the Tropic of Capricorn (subtropical areas). In similar areas, for instance in São Paulo and Paraná States, the temperatures are subject to major fluctuations, including occasional frosts, which can destroy the coffee trees (Table 3.3). In some areas, rainfall is deficient too and in several areas it is not adequately distributed; with either a too short or a too long dry period. Mulching, which is practised in areas of low rainfall in other production countries is not usual in the main production areas of Brazil, since mulched coffee is more vulnerable to frosts. Suitable climatic zones for robusta coffee are found along the coast, in the South-East Region, and in certain parts of the Mid-West and North Regions.

Although often susceptible to erosion, many soils in the major production areas are suitable for cultivation of coffee. The best soil type for coffee in São Paulo State is called 'terra roxa', which is a soil of volcanic origin, clayey but friable in texture, and dark red. It has an excellent physical structure. Other major soils in that state are 'massape', a slightly acid clayey loam and 'arenito', a slightly acid sandy loam, both with a good original fertility and a reasonable physical structure. In the North Region in Rondônia State, coffee grows well on eutrophic red yellow podzols, which are deep well drained loamy soils, with a moderate to high fertility.

3.3 The coffee sector

3.3.1 History

After the discovery of Brazil in 1500 by the Portuguese Cabral, the economic exploitation of the country depended on the large-scale production of a few exports. Brazil has been named after one of those, the tropical wood 'pau brasil' from which red paint was produced. In the 16th and 17th Centuries, sugar was the main export, which was mainly produced along the coast in the North-East Region. The rubber boom in the Amazon Region followed and gold-mining was a major activity in several areas in the 18th Century. In that period, the present states of Minas Gerais, Goias and São Paulo gained importance, and in the 19th Century these states became the major coffee zones.

The first coffee seeds arrived in Brazil in 1727. In that year, the wife of the Governor of French Guyana gave some coffee berries to an admirer, Sergeant Major Francisco de Melo Palheta, who was on an official mission to that French colony. At first, it was grown in home gardens in the northern States of Para and Maranhao, and the average annual amounts exported (from Belem) between 1756 and 1777 were only about 4000 arrobas (50 t), which was 10 % of the amounts of cocoa beans exported in that period. Cultivation of coffee gradually moved southward. When (gold)mining had gone into decline at the end of the 18th Century, and investors were looking for new opportunities, they found those in the coffee trade. The United States of America had just won its independence from Britain and became a major new market for coffee.

The first large coffee plantations were in the Paraíba Valley in the Province of Rio de Janeiro. These were latifundia (large farms) operated with slave labour under the supervision of foremen. By 1810, Brazil was producing 40 % of the total world supply of coffee. Exports increased rapidly, from 18 000 t in the 1820s, 300 000 t in 1881-1890 to about 700 000 t in 1899-1902. In that last period, Brazil accounted for about 70 % of the world's coffee.

Meanwhile, São Paulo had become the major coffee region and this was accentuated at the end of the 18th Century by abolition of slavery and the excessive erosion in the Paraíba Valley. The planter elite, in São Paulo called the Paulistas, had solved two major problems, without having to rely completely on foreign assistance. Their first concern had been to construct a railway network. The Santos-Jundiaí railway was financed in London but its building was organized by a group of Paulistas led by the 'Baron' Maua. From this main line, four trunks were built by different groups of planters to potential producing areas of the interior. Practically all regions in São Paulo State derive their names from the railway companies. The railways and the coffee plantations extended simultaneously across the State.

More difficult to solve was the scarcity of labour caused by the abolition of the slave trade. Agricultural exploitation had been based on slavery for about three centuries. Though the plantations in Paraiba had been operated completely with slaves and profits had been used for new slave purchases until the end of the slavery period, the coffee barons in São Paulo, where slaves had been scarce, realized that it would be better to convert rapidly to a free labour system lest the trade be disrupted. They formed companies to bring in Italian peasants and devised a system of short-term wage contracts. The plantation families established partnerships, originally for the development of the railways, the banking system, the brokerage and the import activities, but later also for manufacturing ventures and agro-industries, such as the cotton spinning and weaving industry, meat-packing plants, maize and manioc mills. Because land was cheap relative to capital and labour, no effort was made to maintain soil fertility and so the 'pattern of a hollow frontier' developed: the plantations gradually moved westwards, leaving behind land fit only for pastures. To develop lands beyond the settled areas required promotion of real estate.

At the start of the 20th Century, the market could no longer absorb the fast-increasing production, and prices fell. In 1902, the first-ever meeting of coffee-producing countries was held in New York between Brazil, Columbia and Cuba, to fix a common strategy for increasing prices. It was decided to limit production and further planting was prohibited. However production went up again and, under the Taubate Convention, which provided for guaranteed minimum prices for producers and for guidelines on quality, the State of São Paulo bought part of the surplus production.

Similar problems occurred in 1914 and 1929. Coffee sales accounted for 72 % of the total value of Brazilian exports and 8 % of GDP in 1929. The world depression then caused the collapse of the Brazilian coffee market in the United States. Prices on the world market were below the cost of production and the newly created National Coffee Department had to buy the coffee at high prices and to store enormous stocks of unsold coffee. It was later decided to destroy these stocks and during the 1930s a total of 78 million bags of coffee were destroyed.

After World War II, production increased rapidly, with rising prices on the world market, which led again to conditions of surplus production and depressed prices in the late 1950s. IBC then carried out an extensive programme of eradication and succeeded in reducing the total number of coffee trees in Brazil from 4300 million in 1960 to about 2300 million by 1967.

3.3.2 Recent developments

Development of coffee production in Brazil in the 1970s has been influenced largely by a severe frost in July 1975, which practically eliminated the 1976/1977 coffee crop in Paraná State. That frost also damaged a large proportion of trees in the other major coffee-producing state, São Paulo, as well as in Mato Grosso State. Little damage occurred then in Minas Gerais State, which rapidly increased its share in total number of trees, but subsequent minor frosts in August 1978 and May 1979 also affected production in that State. Another severe frost occurred in July 1981.

Because of this frost damage, the annual national production fluctuated considerably during the period 1971-1982 (Table 3.4). Domestic consumption was hardly affected, and thanks to the available stocks, exports fluctuated less than production. Because of the large share of Brazil in the total exports of coffee in the world and the relatively low price-elasticity of demand for coffee, prices on the world market rose sharply after the 1975 frost and Brazilian's export earnings of coffee were, surprisingly, higher after than before the frost as a result of them.

Since 1975, production of robusta coffee has increased rapidly and the share in exports of soluble coffee in total coffee exports gradually increased in the 1970s from about 6 % in 1970 to about 20 % in 1980.

Table 3.5 shows the development of area planted and total number of trees by

Table 3.4. Area planted, national production and average yield (1960-1983).

Crop year	Area planted (1000 ha)	Production (1000 t)		Yield (kg/ha)
		arabica	robusta	
1960/61	4893	1788	–	365
1969/70	2730	1260	–	462
1970/71	2565	600	–	257
1971/72	2593	1476	–	569
1972/73	2476	1470	–	594
1973/74	2771	858	–	320
1974/75	2837	1650	–	582
1975/76	2700	1368	12	511
1976/77	2340	540	18	238
1977/78	2742	1026	24	383
1978/79	2867	1146	54	419
1979/80	2980	1254	66	443
1980/81	3020	1215	75	427
1981/82	3145	1890	90	630
1982/83	3225	975	90	330
1983/84	3225	1713	130	571

Source: USDA (1983)

Table 3.5. Area planted (× 1000 ha) and number of trees (× 1 million), by state in three crop years. Proportion (in %) of total area of the state between brackets.

State	1960/61		1970/71		1979/80	
	area	number	area	number	area	number
São Paulo	1392 (28)	1155	828 (32)	687	962 (32)	955
Paraná	1795 (37)	1287	1109 (43)	816	827 (28)	824
Minas Gerais	741 (15)	760	296 (12)	344	581 (20)	877
Espírito Santo	424 (9)	566	204 (8)	254	344 (12)	438
Other states	556 (11)	557	128 (5)	128	245 (8)	300
Total	4908	4325	2565	2228	2959	3394
Trees per ha	881		869		1147	

Sources: USDA (1979), IBC (1973)

state. By the coffee-tree eradication programme undertaken during the 1960s, both area and production of coffee were extremely low in 1966-1973 and stocks diminished rapidly, accentuating the impact of the 1975 frost. The eradication programme had affected Paraná State least, enabling the state to improve its position as major coffee-producing region, but it lost that position after the 1975 frost. Because of that disaster, farmers shifted into soya.

3.3.3 *The organization*

Coffee in Brazil is now grown on about 300 000 farms with an average size of about 60 ha, of which 15 % is devoted to coffee production. Most of Brazil's coffee is grown without shade, harvested by stripping and subsequently processed by the dry method (unwashed arabicas). Wet processing is, however, expanding considerably. Increasingly, producer cooperatives are becoming involved in coffee processing and marketing activities, and even in export. Major participants in the internal market are the millers, who buy from producers and their cooperatives, and sell the green coffee to domestic roasters, the soluble industry and exporters.

Whereas policies in coffee are set by the National Monetary Council, implementation of policies and formulation of programmes is left to the Brazilian Coffee Institute (IBC). A major task of this Institute, created in 1950 as an autono-

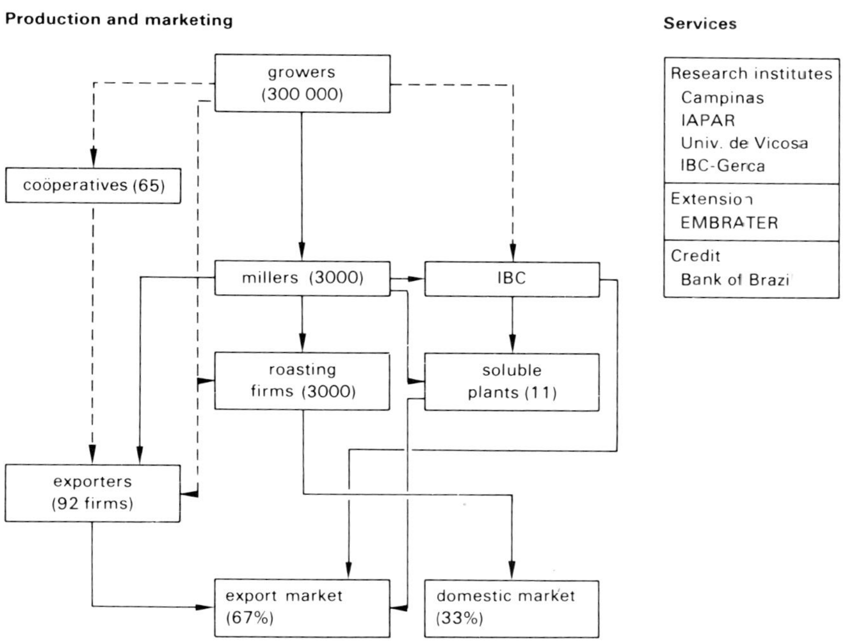

Fig. 3.5. Participants in the coffee sector and main marketing channels in Brazil in 1981. Solid line, main channel; broken line, secondary channel.

mous agency within the Ministry of Trade and Industry, is to maintain the balance between Brazil's coffee production, domestic consumption and exports. It provides growers with a support price, extension services and credit facilities, channelled through the Bank of Brazil.

At various times, IBC buys and then sells from stocks on the domestic and world markets, and now maintains a domestic warehousing network with a capacity of about 60 million bags. It also owns warehouses in several places in Europe and Asia. During the 1960s, IBC successfully implemented a programme of eradication of coffee plants through a scheme of generous credits, subsidies and technical advice for the uprooting of coffee trees and the development of alternative enterprises. After the frost periods of 1969 and 1975 (1975 Emergency Plan), it carried out replanting programmes with similar incentives. Figure 3.5 shows the participants in the coffee sector, and the main marketing channels.

3.4 Production

3.4.1 Production units and farming systems

The large differences in ecological conditions, infrastructure, population density and socio-economic conditions have all contributed to the diversity of agricultural production systems and production units in Brazil. Besides extensive forestry (including production of rubber and Brazil nuts) in the Amazon Region and traditional and modern extensive livestock production in the less populated areas, five broad farming systems can be distinguished.

In tropical rain forest areas in the North Region with ample land, there is a form of shifting cultivation (roça agriculture), with simple tools and mainly subsistence crops. After a few years of cultivation, the poor soils in these areas become exhausted and then require a fallow period of 10 to 20 years. In some of these areas and in the Mid-West, coffee (arabica and robusta), cocoa and rubber are becoming increasingly popular cash crops. In more densely populated areas along the northeastern coast, in Espírito Santo and in the South, these extensive systems made way for sedentary forms of traditional agriculture. The main crop in these traditional farming systems is cassava.

The farmings systems more oriented to cash crops are associated with the following 3 farm types:

– small commercial family-operated farms (sitios),which are important producers of maize, rice, coffee, cocoa and cotton, and are situated relatively close to urban centres

– large commercial farms (fazendas), most of which are found in São Paulo, and Paraná States and which concentrate on capital-intensive production of coffee, sugar-cane, soya, wheat and cotton

– large estates, which have been involved in the capital extensive cultivation of sugar-cane along the northeastern and eastern coast for centuries.

Apart from the size of farm, small and large commercial farms often differ in the degree of diversification and in the type of labour employed, in sitios mainly family labour and on fazendas hired workers, sharecroppers and day workers. Table 3.6 gives information on the production structure in the regions and on the relative importance of the main crops. These data illustrate how farm types and cropping patterns are distributed over the main regions. Extensive farming systems with emphasis on food crops are found mainly in the North and Mid-West Regions. The major industrial crops in the other regions are cotton (in the North-East), coffee and sugar-cane (in the South-East) and soya (in the South). The average farm size is least in the North-East and South Regions, but these averages conceal a considerable disparity between the small and very small farms (minifundios) and the large and very large farms (latifundios).

The distribution of land ownership is extremely uneven: more than 70 % of farms (those with less than 30 ha) together own only 10 % of the land, whereas the large farms (with more than 500 ha), which constitute less than 2 % the total number of farms, own about half the land. Farm size is commonly expressed in alqueires, equal to 2.4 ha. For land reform, classification of farm types by size is

Table 3.6. Regional agricultural production structure (1970) and the importance of the principal crops by area harvested[1] (1977).

	Region					
	North	North east	South east	South	Mid-West	total
Number of farms (million)	0.3	2.2	0.9	1.3	0.3	5.0
Average farm size (ha)	86	33	75	36	320	59
Agricultural work force (million)	1.0	7.8	4.1	4.3	1.0	18.2
Relative importance of principal crops (%)						
Foodcrops	83	56	66	61	85	64
rice	29	8	11	7	50	13
corn	17	19	31	26	24	25
cassava	26	10	2	2	1	5
others	11	19	22	26	10	21
Industrial crops	17	44	34	39	15	36
cotton	–	25	4	2	3	9
sugar-cane	–	7	12	–	–	5
coffee	–	–	12	3	–	4
soyabeans	–	–	5	32	10	15
others	17	12	1	2	2	3

1. Total area harvested is only about 16 % of total farm area, which includes pastureland and land unused.

Sources: World Bank (1982), van der Pluym (1978)

Table 3.7. Avarage number of coffee farms, production of dried cherries and labour involved, by farm size (State of São Paulo, 1973-1975).

Farm size (ha)	Number of farms		Production		Work force	
	(× 1000)	(%)	(1000 t)	(%)	(× 1000)	(%)
3-30	36	52	384	26	116	27
30-100	22	31	388	27	123	29
100-300	8	11	313	22	87	20
300-1000	3	5	264	18	69	16
more than 1000	1	1	105	7	32	8
Total	70	100	1454	100	427	100

Source: Goldenberg et al. (1979)

based on 'model farms', which differ for each agricultural zone. The cropping intensity is inversely related to farm size and ranges from about 90 % for small farms, around 50 % for all farms up to 30 ha to only 5 % for farms with more than 400 ha (van der Pluym, 1978).

About 6 % of all farms in Brazil cultivate some coffee, with about 4 % of the total area harvested (Table 3.6). But in the four major coffee-producing states, coffee ranks third in area behind maize and soya. In the State of Espírito Santo, it ranks first. In those states, coffee employs about 15 % of the labour engaged in agriculture. Most of the farms with coffee fall into the categories of small and large commercial farms. A typical small coffee farm (sitio) has about 10-30 ha, of which 30 to 50 % is under coffee (virtually the only cash crop). By contrast, a large coffee farm (fazenda) has at least 300 ha, with only some 10 % of the land under coffee. Those large farms are often diversified, with some 25 % of the land under arable crops such as maize, soya, upland rice and sugar-cane, some forest or waste land and large pasture grounds for their dairy or beef cattle. Table 3.7 shows data on coffee cultivation in São Paulo State. About 52 % of the farms were less than 30 ha. Those together produced the same amount of coffee as about 4000 farms (6 % of total) with more than 300 ha.

3.4.2 Production stages and production costs

Brazil and Ethiopia are the only countries that predominantly produce unwashed (dry-processed) arabica coffee and both countries practise relatively extensive production systems. In the Ethiopian highlands, coffee grows spontaneously under the original forest, and farmers have only to collect the cherries from the ground, dry them and sell them. By contrast, the extensive production systems in Brazil result from ample land, scarcity of cheap labour and ecological conditions that lead to a short harvesting period.

In South Brazil, coffee production used to be a soil-exhausting activity. After a certain period, the plantations were abandoned, the land went into extensive grazing, and new land was cleared and planted with coffee (Ruthenberg, 1980). To reduce effort on labour, land preparation, spraying and weeding, operations are now more and more mechanized, pruning is hardly applied and harvesting is by stripping.

The best information on the cost of production of coffee is that published by the Institute of Agricultural Economics of the Ministry of Agriculture of the State of São Paulo, which is based on surveys in that state. The data on the establishment costs per hectare (Table 3.8) are derived from studies by that institute in 1976/1977, and are updated to present the situation in 1983. The most common spacing is 4 m × 2.5 m, which gives a planting density of 1000 trees per hectare. In practice, planting density varies, depending on numerous factors, and the costs of production are therefore usually calculated per 1000 trees.

Since the killer frost of 1975, small-scale coffee farmers in Paraná State have preferred to leave more space between the coffee trees (lower density) and interplant them with such crops as maize, soya, rice and castor oil. But with the rising cost of land, new plantings in Brazil have had an average density of around 1400 trees per hectare since the late 1970s. Where coffee is still planted on virgin soil (after the removal of forests), direct sowing is practised with several trees being allowed to grow to maturity in each planting hole. But often, certainly in São Paulo State, new coffee plantations are established on land cleared long before, and transplanting of seedlings is then the usual practice (with often two seedlings per hole).

Until the 1960s, the main varieties were Bourbon, Comum and Maragogipe, but since then the vigorous and high-yielding variety Mundo Novo has become the most popular. Because of its smaller size, the Caturra variety requires a higher planting density and a more intensive production system; it has not (yet) been used widespread.

Table 3.8 shows that during planting mono-superphosphate is applied at a rate of some 300 kg/ha and that gradually increasing amounts of nitrogen and potash are applied during the first years. The total costs of labour, machinery, materials and other inputs for the establishment of 1 ha of arabica coffee amounted in 1983/1984 to about Cr$0.6 million.

A study in depth on production systems and annual production costs in São Paulo State was undertaken in 1958 by the Joint ECLA/FAO Agriculture Division. In this study, a major distinction was made between diversified farms (fazendas only) and specialized farms (sitios and fazendas of different size). Techniques in coffee growing throughout the range of farm size covered seemed to be uniform, with cultivation standards generally low. About 90 % of labour inputs (between 67 to 96 man-days per hectare) were used for harvesting and weeding operations, almost equally divided between the two, and no farm used more than 3 tractor-days and or 13 animal-days per hectare of mature coffee. Seasonal labour require-

Table 3.8. Establishment costs (years 1, 2 and 3) per hectare arabica coffee (1983/84). Variety Mundo Novo; spacing 4 m × 2.5 m; two plants per hole.

Cost elements	1	2	3	Total
Labour inputs (man-days)[1]				
land preparation & lining out	8	–	–	8
digging of planting holes	20	–	–	20
planting	15	5	–	20
weeding & cleaning	5	21	21	47
fertilizing & manuring	6	2	2	10
spraying & rust control	–	–	3	3
total	54	28	26	108
Subtotal (Cr$1000)	135	70	65	270
Tractor inputs (days)				
landpreparation	3.5	–	–	3.5
transport	0.3	0.1	0.2	0.6
spraying	–	–	(0.8)	(0.8)
total	3.8	0.1	1.0	4.9
Subtotal (Cr$1000)	76	2	20	98
Material and other inputs (quantities)				
plants (Cr$40 per plant)	2000	200	–	2200
fertilizers: ammonium sulphate (Cr$110/kg)	60	160	300	520
superphosphate (Cr$105/kg)	300	–	–	300
potassium chloride (Cr$120/kg)	40	80	80	200
spraying copper-oxychloride (Cr$2500/kg)	–	–	12	12
animal traction (days, Cr$300/day)	–	–	5	5
Subtotal (Cr$1000)	123	35	74	232
Total establishment costs (Cr$1000)	334	107	159	600

1. In forested areas (e.g. Rondônia) also including 36 man-days of manual land clearing (and chainsaw running costs).
Source: Instituto de Economia Agricola (1976, 1981 & 1983; extrapolated)

ments reached their peak in the main harvest months of June, July and August. A work force 15-70 % larger was needed during these months. The costs of fertilizers and manure were 15-30 % of the costs of labour. Since the costs of family labour were not calculated, this proportion does not apply for sitios, which used a considerable amount of manure. The average coffee yield for the State as a whole was about 450 kg/ha in 1958 and, at the then prices, coffee constituted the most profitable activity, apart from sugar-cane and onions (ECLA/FAO, 1960).

How does the situation then compares with the situation, 25 years later, in 1983/84? Table 3.9 gives the input requirements and production costs, estimated for that year by the Institute of Agricultural Economics of São Paulo State, for three major production areas in that state: Ribeirão Prêto, Campinas and Bauru. For comparison, it also provides estimates from other sources for production zones in two other states.

The data provided for the three areas in São Paulo State show what changes had occurred since 1958. Because of the increased use of herbicides, less labour was used in 1983/1984 for weeding, which, together with sweeping, accounted for about 45 % of all labour inputs in 1958. Sweeping includes two operations, arruaçao and esparramçao, in which leaves are swept away from the tree before picking and swept back under the tree afterwards. Fertilization has become more important, with respective average rates of N, P_2O_5 and K_2O of about 180, 40 and 160 kg/ha in 1958. Generally recommended rates are 300, 150 and 300 kg/ha. These rates exclude manure, of which considerable amounts are applied (2-7 t/ha), and minor elements.

A major new activity is spraying with copper-based fungicides, which has become widespread since the outbreak of coffee rust (*Hemileia vastatrix*) in the late 1960s. To control rust effectively, 4 to 7 sprayings are needed each year, using copper oxychloride (50 %) at a rate of 4-6 kg/ha for each application. Animal traction has become less important and the average number of tractor-days has increased with the increased use of herbicides, fertilizers and fungicides, whose application is often mechanized. A major and unpredictable cost is interest on bank loans, in recent years always rising, with the high rate of inflation (since 1982 over 100 %).

The other two states for which estimates of annual costs of production are given in Table 3.9 have more favourable ecological conditions for robusta than for arabica (lower altitude and higher temperatures). Espírito Santo traditionally produces considerable amounts of robusta, and Rondônia is considered as a potential robusta-growing area. Production costs and yields in these two areas (Table 3.9) refer, however, to arabica coffee. Labour inputs in these areas exceed those in São Paulo, mainly because of the higher labour requirements for weeding operations. Otherwise the costs of production in Espírito Santo are similar to those in São Paulo State. The production system in Rondônia is illustrative for cultivation methods used in the past in all coffee areas. After clearing the forested land, coffee is produced without any fertilizer until the soil is exhausted and the plantation abandoned. Costs of establishment include manual land clearance and annual costs include only few material inputs and no power inputs. Yields are considerable and costs per kilogram of coffee produced are only 60 % of those in São Paulo State.

Robusta requires less disease control (immune to leaf-rust) but more labour for picking. Compared with arabica, its total costs of production are probably somewhat lower in Espírito Santo and about the same in Rondônia State and average yields are slightly higher in both areas. But farmers in Rondônia prefer to grow arabica because of the selective picking required for robusta, due to sequential maturing of the berries. They are, however, also more familiar with arabica.

Table 3.9. Production costs of green coffee (per hectare and per kilogram) for different states and regions (1983/84).

	São Paulo			Espírito Santo	Rondônia
	Ribeirão Prêto	Campinas	Bauru		
Yield (kg/ha)	800	730	600	600	800
Plant density (plants/ha)	1000	1000	1000	1000	1000
Wage rate (Cr$/man-day)	2800	2600	2400	2400	3000
Rainfall (mm/year)	1350	1200	1150	1400	2000
Soil type[1]	TR	AR/TR	AR	.	RP
Altitude (m)	550	550	450	300	250
Annuity of establishment costs (Cr$ 1000) (10 %, 20 years)	70	70	70	70	80
Annual costs					
Labour inputs[2] (man-days)					
weeding	18	15	12	34	30
sweeping	6	8	5	–	4
fertilizing	7	5	4	4	–
disease control	3	4	2	11	3
transport	1	1	1	–	2
harvesting	36	35	32	25	40
processing & other activities	3	4	6	14	5
total	74	72	62	88	84
Subtotal of costs (Cr$ 1000)	207	187	149	211	252
Power inputs (days)					
tractor[3]	4.7	2.3	3.3	4.0	–
animal traction	–	1.0	–	–	–
motor spraying	–	3.6	2.2	–	–
Subtotal (Cr$ 1000)	97	57	74	80	0
Material inputs (Cr$ 1000)					
fertilizers & herbicides	143	134	87	114	–
manure	43	59	22	–	5
insecticides & fungicides	118	52	59	35	10
processing materials & bags	30	27	23	23	30
Subtotal	334	272	191	172	45
Overhead costs (Cr$ 1000)					
administration (5 % of variable costs)	32	26	21	23	15
depreciation & interest on capital	34	17	23	20	10
interest on loans	284	224	176	202	160
Subtotal	350	267	220	245	185
Costs per hectare (Cr$ 1000)	1058	853	704	778	562
Costs per kilogram (Cr$ 1000)	1.32	1.17	1.17	1.30	0.70
(US$)	1.29	1.15	1.15	1.27	0.69

1. TR = Terra Roxa, AR = Arenito; RP = Red podzols.
2. Includes tractor driver.
3. Includes operational costs of equipment.

Sources: Instituto de Economia Agricola (1983), Ruthenberg (1980), FAO/Investment Centre (1980)

3.4.3 *Yields and returns*

The average yields for 5 production areas (Table 3.9) are somewhat higher than the national average yield in recent years of about 500-650 kg/ha, but the structure of production costs is illustrative for the whole of Brazil. Growers generally sell their coffee to millers and traders in 40-kg bags of dry cherries, equivalent to a third of a 60-kg bag of green coffee.

Prices to growers vary between states (depending on quality) and are influenced by the guaranteed minimum price offered by the Government, through IBC (Section 3.6.3). There are two major quality classes for arabica: the higher quality coffee from Paraná, São Paulo and Sul de Minas (Minais Gerais) falls within Group I, and the lower-quality coffee from Espírito Santo and Zona da Mata (also Minais Gerais) in Group II. Because of the rapid inflation in recent years, the producer prices increase constantly and the guaranteed prices are often adjusted upwards during the year.

In June 1982, the producer prices were Cr$286 and Cr$265 per kilogram (green coffee equivalent) for Groups I and II, respectively, and at the end of the 1982/83 season, the price for Group I stood at Cr$617 per kilogram of green coffee. The costs of production in São Paulo State for that season were Cr$617, Cr$389 and Cr$601 per kilogram green coffee, for the Ribeirão Prêto, Campinas and Bauru areas, respectively. The high net returns for the Campinas area were due to the good yields obtained of 800 kg/ha, against 600 and 400 kg/ha for the Ribeirão Prêto and Bauru areas, respectively.

Since the coffee in Espírito Santo is classified into Group II, producer prices are generally lower than in São Paulo State and they probably did not cover the integral costs of production in 1982/1983. Small coffee farmers in Paraná State have had difficult times since the 1975 frost, because of the low yields in the years after that frost, the large investments required and the high interest on loans paid for re-establishment of their plantations (Nogueira da Gama, 1979).

3.5 Processing

3.5.1 *Milling and roasting*

For the coffee exported as green beans, processing is of minor importance. The farm processing generally consists of drying only, although wet processing is increasingly applied. Hulling is the only major operation and is undertaken by about 3000 millers in the production areas. Milling charges amounted in 1979/1980 to about Cr$80 for a 60-kg bag of green coffee, or about 2.5 % of the net minimum guaranteed price to producers.

Over the years, the millers have increased their activities and now play a leading role in the internal marketing of coffee (Section 3.6). Together, they bulk, process, grade and sell about 85-90 % of all the coffee produced in Brazil, or more than 1.6 million tonnes in a peak year (e.g. 1981/1982). So on average, a miller handles 300-550 t per year and his capacity should be tuned to the higher figure. The demand for milling services is highly seasonal, the bulk of the dry cherry coffee being delivered in August to October. Through the provision of additional credit for holding stocks on the farm, the Government tries to smooth out the sales of cherry coffee.

The coffee-manufacturing industry, made up of roasters and soluble processors, is important in Brazil. Because of the large population and the relatively high consumption of coffee per person, there is a large domestic market for coffee. So the roasting industry is large, with about 3000 independent firms.

3.5.2 Soluble-coffee industry

Brazil's most spectacular, though also controversial, method of increasing earnings from coffee has been through exports of soluble or instant coffee. These exports increased from 10 000 t green coffee equivalent (or less than 1 % of total coffee production) in 1965/1966 to 200 000 t (10 % of production) in 1981/1982. In the period 1978 to 1983, exports of soluble coffee made up 15 % of total exports of coffee.

There are 11 soluble-coffee factories in Brazil, of which 3 produce freeze-dried instant coffee and the others spray-dried powder. Together, they have an annual capacity of about 80 000 t of instant coffee. In 1975, only 8 such factories were in operation, the total capital invested in them was about US$42 million and turnover was about US$93 million. These factories then employed 2800 workers or 400 per plant. Four of these factories were then foreign-owned, two by Japanese firms (1975 data on soluble-coffee factories from Sampaio, 1980). Brazil is also involved in the construction of complete factories for export to other countries.

Although a major part of the robusta coffee produced in Brazil is used for processing into soluble coffee, the proportion of robusta in soluble blends averages only about 20 %, the remainder being made up of the lower-quality arabica coffee. So Brazilian instant coffee is superior to that produced in the United States and Europe, made almost entirely from robusta coffee. The extraction rate is, however, lower for arabica coffee. With a uniform conversion rate between soluble and green coffee of 0.385, as applied by the International Coffee Agreement for the calculation of export quotas, Brazil uses somewhat more coffee to fill that part of its quota that consists of soluble coffee. Because of the Brazilian soluble exports, the United States became a net importer instead of a major exporter of instant coffee in the late 1960s. In 1968, the United States pressed Brazil into imposing a 10 % tax (additio) on soluble coffee, on the argument of unfair competition and this tax became effective as from 1 May 1969. Until the early 1970s, IBC in fact supplied the green coffee requirements of the industry at about 50 % below world prices. To ensure adequate supplies at low cost, to soluble processors and roasters, the export of low-quality green coffee is prohibited.

3.6 Marketing

3.6.1 Marketing channels and functions

A major role in the marketing of coffee in Brazil is played by the millers, who, through agents and traders, directly or indirectly receive 80-90 % of coffee sold by growers, usually in 40-kg bags of dried cherries. A minor part of the coffee production (in 1978/1979 12 %) is milled by growers and their cooperatives and sold directly to exporters and the manufacturing industry. Since the harvesting season is generally short and uniform over the major production areas (June to August), prices offered by millers tend to be somewhat lower during the peak period, and farmers and traders that can afford to, retain and store the dried cherry coffee in anticipation of better prices.

After milling and grading, the millers sell the green coffee to the major participants in the export trade: IBC, the soluble industry, private exporters, intermediaries, and to roasters supplying the domestic market. Of the volume handled by millers in 1978/1979, about 40 % was purchased by IBC, 20 % by exporters, 10 % by the soluble-coffee industry and 30 % by roasters. The extent to which millers sell coffee to exporters or to IBC is largely determined by Government manipulation of the various taxes and prices decreed by IBC.

The coffee-marketing season starts on 1 July and new policies for the coming year are generally announced in June. A major instrument is the minimum guaranteed price, paid to growers, for green coffee delivered to the IBC warehouses. It varies a little between areas and with quality. It is generally fixed at one level for several months, and at a slightly higher level for the period thereafter, to encourage growers to store some of their coffee. However the high inflation influences such decisions.

When world coffee prices are relatively stable, the guaranteed prices are generally close to payments to growers by private traders. But at times of rising world prices, the private sector usually offers higher prices and only small amounts are bought by IBC.

The major marketing functions, exercised during history by private trading companies, State Authorities and the National Coffee Institute (IBC), were those of storage and financing of the annual coffee crop to influence or stabilize both domestic and world prices. For that purpose, IBC operates a network of warehouses, with a total capacity of about 3 million tonnes, which is equivalent to one and a half to two years of production. The total national storage capacity for coffee reached 7.2 million tonnes in 1984, distributed over about 1300 warehouses. The public (IBC) warehouses are among the largest and more than half of them are in Paraná State.

The importance of transport has already been illustrated in Section 3.3.1 (history) for Paulo State. The further the production areas moved from the coastal area, the higher the transport costs. The produce of such remoted areas as Rondônia State is therefore intended for the domestic market rather than for export.

Since Brazil always has been the major producing country, the classification system used for Brazilian coffee has been adopted in the trade and in other production countries. It is generally known as the 'type and description' system, in which 'type' in fact only refers to the number of defects in a certain amount of coffee and 'description' covers the other characteristics. The main criteria for classification are given below with a few examples.

1. Number of defects (e.g. black or broken beans, stones, shells). The Brazilian system ranges from Type 2, 4 defects in 300 g to Type 8, 360 defects per 300 g. The New York system ranges from Grade 2, 6 defects per 36 cubic inches, to Grade 7, 240 defects.
2. Botanical species, e.g. arabica, and cultivar, e.g. Maragogype.
3. Geographic origin by port, e.g. Santos, and by state, e.g. São Paulo.
4. Size of bean, ranging from screen 20 (large) to screen 13 (small).
5. Colour of bean, e.g. green, greenish, yellowish (usually older).
6. Taste: soft (almost absent because of fertilizers), hard and Rio (harsh taste of the State of Rio and other areas exposed to ocean winds).
7. Shape, e.g. flat, round (peaberry).
8. Processing method: usually dry but now increasingly wet method. Other criteria are those by roast, by year of harvesting, by density and by specific types of exporters. Now samples can be sent by air mail in order to establish contracts but the same terminology is still used.

3.6.2 Pricing, taxation and marketing margins

To fulfill the objectives of maximizing export earnings, supplying the domestic market, and ensuring an adequate return to producers and reasonable margins to traders and exporters, the Government (through IBC) has developed a set of minimum prices and taxes, through manipulation of which it controls the marketing system. Apart from the guaranteed minimum price to growers, a gross price from which ICM (a value-added tax), a local tax and the milling charges have to be deducted to obtain the real price to producers, there is also a minimum export price or registration price. In 1980, ICM tax rates were 11-15 %. About 80 % of the tax revenues accrue to the Government and 20 % to local authorities.

All sales by exporters have to be registered with IBC, which regulates them, also to the international quota allocation. IBC only accepts registrations at or above the minimum registration price. But the final export price is mainly determined by the level of the 'confiscao' or 'contribution quota', which is the second element of the registration price. This is a kind of tax, in the form of a fixed levy per 60-kg bag, which exporters have to pay (in dollars) to the authorities from the proceeds of their exports. The revenues of this tax are partly used for programmes related to the coffee sector.

So the exporters have to offer prices to producers at about or above the guar-

Table 3.10. Prices, marketing costs and margins in Cr$ and US$ per kilogram[1] (green coffee equivalents) in three years.

	1974		1979		1980	
	US$	Cr$	US$	Cr$	US$	Cr$
Export unit value (f.o.b.)	1.24	8.40	3.41	91.90	2.74	1960
Costs and margins						
Contribution quota	0.24	1.60	1.42	38.25	1.44	1027
handling, transport, port charges	0.04	0.30	0.07	1.80	0.06	43
margin of exporter	0.04	0.30	0.07	1.94	0.06	44
local tax (ICM)	0.09	0.60	0.20	5.43	0.17	117
rural fund tax	0.02	0.13	0.04	1.04	0.02	17
milling charges	0.02	0.13	0.04	1.00	0.02	16
new bags and other costs	0.01	0.09	0.02	0.70	0.02	13
Total	0.46	3.15	1.86	50.16	1.79	1277
Price to producer	0.77	5.25	1.55	41.74	0.95	683
Producer price as share of export unit value (%)	62		45		35	

1. Foreign exchange rates (Cr$ per US$) respectively 6.79; 26.95 and 716.00.
Source: ICO (1977-1983)

anteed prices, while selling the coffee abroad at prices well above the contribution quota, in order to cover their costs. Since guaranteed prices do not vary between the good and the normal types (up to Type 6) and are only reduced for the lower qualities (Types 7 and 8), the common grades usually go to IBC, and the exporters take the better qualities at a premium in price. Table 3.10 gives the marketing margins and costs for the export of green coffee, in the years 1974, 1979 and 1980.

3.6.3 Export and domestic consumption

With its large population and the traditional importance of coffee in its economy, Brazil is (after the United States) the second largest consuming country in the world, with about 10 % of total world consumption. But with declining production and emphasis on exports, average consumption of coffee per person went down from about 6 kg in 1970 to less than 4 kg in 1980. On average, domestic consumption constitutes about a third of national production (Table 3.11). But in 1976, the year after the last severe frost, annual production only just covered domestic needs. Stocks had fallen from a high level of 3.3 million tonnes in 1968 to a moderate level of about a million tonnes in 1975 and were just sufficient for Brazil to benefit initially from the rising prices in 1976. But with prices at a peak in 1977, stocks were exhausted and exports dropped to a low level (Table 3.11). When the international quota system was reintroduced in 1980, Brazil was still unable to use

Table 3.11. National production, domestic consumption, exports and stocks (in million tonnes). Crop year starting 1 April (1972-1982).

Year	Opening stock[1]	Pro-duc-tion	Con-sump-tion	Available for export	Export		
					total	ICO members	non-members
1972	1.4	1.6	0.5	2.5	1.1	1.0	0.1
1973	1.4	1.0	0.5	1.9	1.2	1.0	0.2
1974	0.7	1.6	0.4	1.9	0.8	0.6	0.2
1975	1.1	1.4	0.4	2.1	0.8	0.7	0.1
1976	1.3	0.4	0.4	1.3	1.1	0.9	0.2
1977	0.2	0.9	0.4	0.7	0.4	0.4	0.0
1978	0.3	1.2	0.4	1.1	0.8	0.7	0.1
1979	0.3	1.3	0.4	1.2	0.8	0.6	0.2
1980	0.4	1.1	0.5	1.0	1.0	0.8	0.2
1981	0.0	1.9	0.4	1.5	1.0	0.9	0.1
1982	0.6	1.1	0.5	1.2	1.0	0.9	0.1

1. Total stocks less working stock of 0.56 million tonnes.
Source: Marshall (1983)

its full allocation, and had to declare a shortfall, to be shared by other countries. The country was forced to do so again in 1981, after the frost in July.

The number of exporters is around 90, of which about 20 have been long in business whereas others come and go. Quotas are apportioned among them by IBC, according to past performance. IBC, often a large buyer, sells mainly on the local market, to roasters, the soluble industry or to exporters. It also exports a certain amount of coffee and ships large amounts to its overseas warehouses in Trieste, Hong Kong and Hamburg.

Table 3.12 shows the destination of Brazilian exports of green and soluble coffee. The United States are by far the largest buyer, followed by Italy and a few other European countries. The United Kingdom is a large buyer of soluble coffee. Spain ranks second to the United States as buyer of robusta coffee from Brazil. Because the associate countries of the European Community in Africa are exempt from a 5 % import tax (imposed by Community countries), robusta exports from those countries have a price advantage over those from Brazil.

Over the past fifty years, the share of Brazil coffee in the world market has declined from over 60 % in 1930 to a mere 20-25 % in 1980. Unwashed arabica coffee from Brazil was inferior to the mild arabicas, but roasters abroad generally used a fair amount of it in their blends. However they could partly substitute low-quality milds or cheaper robusta coffee, the production of which increased rapidly in the 1950s and 1960s.

To face this competition and to guarantee its share in the market, Brazil has since the 1960s operated the 'special deal' system. A roaster or manufacturer

Table 3.12. Proportional distribution (%) of green and soluble coffee exports to major importing countries.

	Green Coffee		Soluble coffee	
	1970-1972	1979-1981	1970-1972	1979-1981
ICO members	88	85	87	98
United States	33	23	49	48
Italy	14	11	1	0
West Germany	5	7	4	6
France	4	6	1	1
Sweden	6	6	0	0
Japan	1	5	0	4
Spain	2	4	0	0
United Kingdom	2	1	22	29
others	21	22	10	10
Non-members	12	15	3	2
Argentina	3	3	.	.
Poland	1	4	.	.
others	8	8	.	.

Source: Marshall (1983)

agrees to buy a certain amount of coffee over a stipulated period in exchange for a discount from the official prices. The discount is not fixed, but varies accordingly to the development of the indicator prices for robusta and other milds. If countries producing such coffee try to undersell Brazil, the discount increases and the market goes down. The discounts are not paid in cash but in the form of credit notes (avisos), which are only issued after considerable delay and then serve for a subsequent purchase. In this way, roasters continue to buy Brazilian coffee to use up the avisos and, since they are committed, they will at times forego purchases from other countries. As a result of them, Rwanda and Burundi were badly affected in 1980, since their crop comes to the market in about the same period as the coffee from Brazil.

3.7 Supporting services

3.7.1 Research and extension services

Because of the ample land available for agricultural expansion, which could be brought into cultivation at low cost, little emphasis used to be paid to development of agricultural research to achieve rapid crop yield increases. But gradually land became scarcer and the need arose to improve agricultural technology. In 1973, the Brazilian Agricultural Research Enterprise (Embrapa) was established together with the Brazilian Enterprise for Technical Assistance (Embrater), both of which fall under the Ministry of Agriculture. Embrapa coordinates eleven national centres for product-specific research, three regional centres and several other specialized research institutions. The major state research institute is that of the Secretariat of Agriculture of the State of São Paulo.

There is not one specific national centre for coffee, but research activities are spread over a few institutions. Agronomic research and breeding activities are done in particular by three institutes in the major coffee states, the 'Instituto Agronomico de Campinas' in Saõ Paulo State, the 'Fundação Instituto Agronomico do Paraná' (IAPAR), and the 'Universidade Federal de Viçosa' in Minas Gerais State. The Institute of Campinas is internationally known. The research undertaken by IBC-Gerca is more practically oriented and directly related to extension activities.

Technical assistance to farmers was a major component of the eradication and replanting programmes in the early 1960s and in the period after the severe frosts of 1969 and 1975 (Section 3.7.2).

3.7.2 Credit and subsidies

The provision of credit and subsidies has played a major role in the Government policies in the coffee sector. Farmers in Brazil always responded rapidly and massively to rising prices on the world market but a few years later Brazil was then faced with a large excess supply and depressed prices. To restore prices, stocks had even to be destroyed (4.7 million tonnes in the period 1931 to 1944).

In the early 1960s, IBC therefore carried out an extensive programme of eradication to reduce the productive capacity. This programme was successful, thanks to the generous credits, subsidies and technical assistance aimed at uprooting coffee trees and developing alternative crops. The number of trees was reduced from about 4.3 thousand million in 1960 to only 2.3 thousand million in 1967. Unfortunately for Brazil, the conditions on the market improved just when the country was hit by a severe frost in 1969.

IBC then embarked on a large replanting scheme, which was further enlarged under the 'Emergency Plan' after the major frost of 1975. Under this plan, about 370 million trees had to be rejuvenated by pruning, another 450 million trees had to be cut right back and about 470 million trees were completely damaged and had to be replanted. It was realized that coffee areas had expanded too much to the south. To redress this trend, the plan aimed to plant another 100 million trees in areas not affected by frost. Under the Emergency Plan, farmers could obtain loans at a subsidized rate of 7 % and repayment periods of 18 months for production of seedlings and of 6 years for planting activities. At that time, the annual rate of inflation amounted to 40 % and commercial loans were provided at a rate of interest of up to 50 %. Table 3.13 shows the impact of the Emergency Plan on the coffee plantations in the states.

Table 3.13. Rehabilitation and replanting under the Emergency Plan, by State (US$1 = Cr $ 8.13) (1975).

Activities	Number of trees (million)				Total	Loansum	
	Paraná	São Paulo	Minas Gerais	others		total (Cr$ million)	per 1000 trees (Cr $ 1000)
Trees to be pruned	88	272	–	9	369	296	0.80
Trees to be stumped	328	120	–	–	448	537	1.20
Replacement	310	100	29	35	474	2843	6.00
New planting	–	20	40	40	100	600	6.00
Total	726	512	69	84	1391	4276	3.07
Undamaged stock	0	293	620	460	1373		
New total stock	726	805	689	544	2764		

Source: ICO (1980)

With another 3800 million cruzeiros for additional activities, the total sum committed by the Government for the Emergency Plan amounted to about Cr$8100 million, or about US$990 million. Apart from the pruning programme, the target for most of the activities was reached within two to three years. To qualify for the loans, farmers had to follow the recommendations of extension officers, for instance in cultivation practices and varieties to be planted. Under the programme, most coffee plantations now, in the 1980s, consist of relatively young trees and of improved varieties ('Mundo Novo'). Credit was also considered the major tool for the promotion of other crops, such as soya beans, wheat and rice, and here again loans were generally made available at low interest rates, taking into account the high rate of inflation.

In 1979, total crop-specific credit amounted to Cr$270 thousand million (US$10 million). Of this total, the coffee sector obtained 11 %, which is high in relation to its share in the total area harvested but slightly lower than its share in the total value of crop production in that year (13 %). More than 20 % of the value of crop loans went to soya bean producers in 1979. Wheat, rice, soya and cotton received a relatively large amount of credit, compared to their contributions to output, whereas cassava received little.

Since 1979/1980, several measures have been taken to raise the effective interest rates on rural credit. A large amount of subsidized credit had reached the large grain and export-crop farmers in the south and south-east, and had pushed up the price of land, the ownership of which is usually a prerequisite for credit.

3.7.3 *Input supply*

The major inputs for coffee production, apart from labour and power, are fertilizers, fungicides, insecticides and herbicides. Use of fertilizer expanded rapidly in

Brazil during the 1970s: from about 1 million tonnes of nutrients in 1970 to about 4 million tonnes in 1980. Lack of phosphorus in the major Brazilian soils and the importance of soya bean production, which requires little nitrogen, mean that the proportions of different fertilizers is roughly as follows: nitrogen 20 %, phosphates 50 % and potassium 30 %. Potassium is entirely imported, whereas nitrogen and phosphorus fertilizers are 40 % and 75 % locally produced, respectively.

By import substitution policies, the domestic fertilizer industry has since the late 1960s enjoyed substantial tariff protection and quota benefits, which together with the continuous high cost of local production contributed to high prices for farmers. But farmers benefited from the subsidized credit for the purchase of fertilizers and they could obtain a 40 % direct subsidy on fertilizer purchases in 1975 and 1976 when prices on the world market of fertilizers had suddenly increased after the oil crisis. Coffee accounts for about 11 % of fertilizer consumption in the country, against 21 % for soya and 15 % for sugar-cane. The amount of fertilizer used in coffee production in 1980 had an estimated value of about US$140 million.

In 1980, the total amount of insecticides, fungicides and herbicides consumed in the country reached about 90 000 t, of which half was produced domestically. The tariffs levied on the import of these chemicals stood at 37 %. The consumption had been somewhat lower in the years after the oil crisis in 1974.

Over the past decade, investment in depulping, washing and modern drying equipment has been promoted to improve coffee quality, but this equipment has mainly been acquired by the large farms.

3.8 Summary and economic parameters

Since 1810, Brazil has been the largest producer of coffee in the world and in the period 1972 to 1982 accounted for 25 % of world production. But changing climate, particularly the occurrence of regular frosts in several production areas has caused annual production to fluctuate considerably in the past ten years. Resulting from these frost periods, Brazil no longer has the bulk of world stocks of coffee. Because of ample land and the relative scarcity of cheap labour, production systems are generally extensive, with increasing emphasis on mechanized operations. The country mainly produces unwashed arabica, but the production of robusta coffee is increasing in the areas that are more suitable for that type of coffee.

Brazil is the only coffee-producing country that manufactures a considerable amount of soluble coffee. The millers perform a central role in the domestic marketing of coffee, with IBC, the National Coffee Institute regulating the market, in particular by setting minimum producer and minimum export prices. These prices are greatly influenced by the prices on the world market and by the level of the 'contribution quota', a kind of tax paid by exporters. Since the coffee boom after the 1975 frost, this contribution quota has been raised to such a level that producers received less than half of the export unit value.

Economic parameters of the coffee sector in Brazil (1983).

Resources used in coffee sector	
Land under (productive) coffee	3.0 million ha
as share of arable & permanent crop land	5 %
Labour involved[1]	1 000 000 man-years
in production	920 000 man-years
in processing, marketing & services	80 000 man-years
share of national work force	2 %
Material inputs for coffee production	Cr$660 000 million*
Annual import requirements for coffee industry	US$270 million*
Income from coffee sector	
Value of coffee production	
(volume × export unit value)	Cr$2 500 000 million
as share of agricultural production	10 %
as share of GDP	1 %
Value of coffee exports (f.o.b.)	US$2400 million
as share of agricultural exports	22 %
as share of exports	10 %
Share of total production value (1979-1983) obtained by	
Producers	45 %
Processing industries, private traders & exporters	10 %
Government and parastatal organizations	45 %
Other parameters	
Unwashed arabica production	93 %
Robusta production	7 %
Yield (green coffee)	570 kg/ha
Production of labour involved (green coffee)	5.7 kg/man-day
Share of coffee produced for export	67 %
Ratio producer price (per kg green coffee):	
daily wage rate (Cr$/man-day)	0.33

1. 1 man-year = 300 man-days.
Sources: FAO (1971-1983), IMF (1983), World Bank (1984), and own estimates

Producers have benefited considerably from generous subsidies and credits at low interest, in particular under the rehabilitation and replanting scheme of the Emergency Plan. Since enough land is still available and progress has been made with mechanization, and because of the modest costs of production and marketing, there is scope for further expansion of coffee production.

References

ECLA/FAO, 1960. Coffee in Latin America; II Brazil, State of São Paulo FAO, Rome.

FAO/Investment Centre, 1980. Preparation report for Rondônia rural development project. FAO, Rome.

FAO, 1971-1983. Production Yearbook. FAO, Rome.

Goldenberg, I.J.E., Ghilardi, A.A. & Topel, R., 1979. Comercializaçao do café no estado de São Paulo ao nivel do produtor. Instituto de Economia Agricola, São Paulo.
Hohé, M., 1979. Landendocumentatie 1978, no. 4, Brazilië. KIT, Amsterdam.
IBC, 1973. Essays on coffee and economic development. IBC, Rio de Janeiro.
IBC, 1980. Coffee in Brazil. IBC, London.
ICO, 1977-1983. Quarterly Statistical Bulletin on Coffee (several issues). ICO, London.
ICO, 1980. Coffee in Brazil. ICO, London.
IMF, 1983. International financial statistics. IMF, Washington.
Instituto de Economia Agricola, 1976, 1981 & 1983. Informaçoes Economicas. São Paulo.
Mc. Farquhar, A.M.M. & Evans, G.B.A., 1972. Employment creation in primary production in less developed countries: Case studies of coffee sectors in Brazil and Kenya. OECD, Paris.
Marshall, C.F., 1983. The world coffee trade. Woodhead-Faulkner, Cambridge.
Matsunaga, M. et al., 1980. A evoluçao da caféicultura em São Paulo. Instituto de Economia Agricola, São Paulo.
Nogueira da Gama, A., 1979. Revista do Comercio de café, Sept. 1979. Radical changes in Parana State's Coffee Culture.
Pluym, Th. van der, 1978. Brazilië. Politieke economische ontwikkeling en ruimtelijke expansie. Amsterdam.
Ruthenberg, H., 1980. Farming systems in the tropics. Clarendon Press, Oxford.
Sampaio, P., 1980. Le capital agro industriel étranger au Brésil. CETRAL, Paris.
Schwerdtfeger, W., 1976. World survey of climatology, Volume 12; Climates of Central and South America. Elsevier, Amsterdam.
USDA, 1979. Foreign agriculture circular. Coffee production and trade in Latin America. USDA, Washington. USDA, 1983. Foreign agriculture circular on coffee. USDA, Washington.
World Bank, 1982. Brazil; A review of agricultural policies. Latin America and the Caribbean Regional Office, Washington.
World Bank, 1984. World development report. Oxford University Press, Washington.

Currency equivalents and price indices in Brazil (1970-1984).

	Exchange rate (Cr$ per US$)	Consumer price index (1980 = 100)
1970	4.6	5
1971	5.3	6
1972	5.9	7
1973	6.1	8
1974	6.8	10
1975	8.1	13
1976	10.7	18
1977	14.1	26
1978	18.1	36
1979	27.0	55
1980	52.7	100
1981	93.1	206
1982	179.5	407
1983	577.0	985
1984	1021.1	1661

Source: IMF (1983)

4 Coffee in Colombia

4.1 Background

Situated in the northwestern part of South America, Colombia with a land area of 1 140 000 km² is the fifth largest country in Latin America after Brazil, Argentina, Mexico and Peru. In 1982, the total population was about 27 million, of which 35 % lives in the four largest cities Bogota, Medellín, Cali and Barranquilla. Administratively, Colombia is divided into 22 departments, 5 'intendencias' and 4 'comisarías' (in the Amazon and Orinoco Areas) and the Bogota Special District covering the capital.

The GNP per person was about US$1460 in 1982. The share of the major sectors in the GDP in that year was agriculture 26 %, industry 32 % and services (including government) 42 %. Livestock accounts for 40 % of the contribution made by the agricultural sector. The sectors contributed as follows to total exployment: agriculture 26, industry 21 and services 53 % (World Bank, 1984). Although many different minerals are found, the mining sector is unimportant. For its energy supplies, Colombia depends on the production of coal, hydro-electricity, oil and gas. However during the 1970s, it became a net importer of petroleum products. The increasing use of firewood and charcoal in rural areas has caused deforestation and severe erosion.

Table 4.1 shows the major crops in Colombia. The production of coffee contributes directly to 5 % of employment and on average to about 7 % of GDP. It is also by far the major export product, with about 60 % of the total value of exports in the period 1976-1980. There are no other major export products, but exports diversified in the 1970s. Other agricultural exports are sugar, cotton, bananas and flowers.

With an average annual production of 748 000 t in the period 1979/81, Colombia contributed 14.5 % of the world production of coffee. It is the second producer in the world, after Brazil.

4.2 Ecology

On the basis of differences in topography and climate as influenced by the three mountain chains of the Andes and other physical and socio-economic factors, the country can be divided into five major regions: the coastal areas along the Caribbean Sea and the Pacific Ocean; the Central Highlands; and the Orinoco Plains

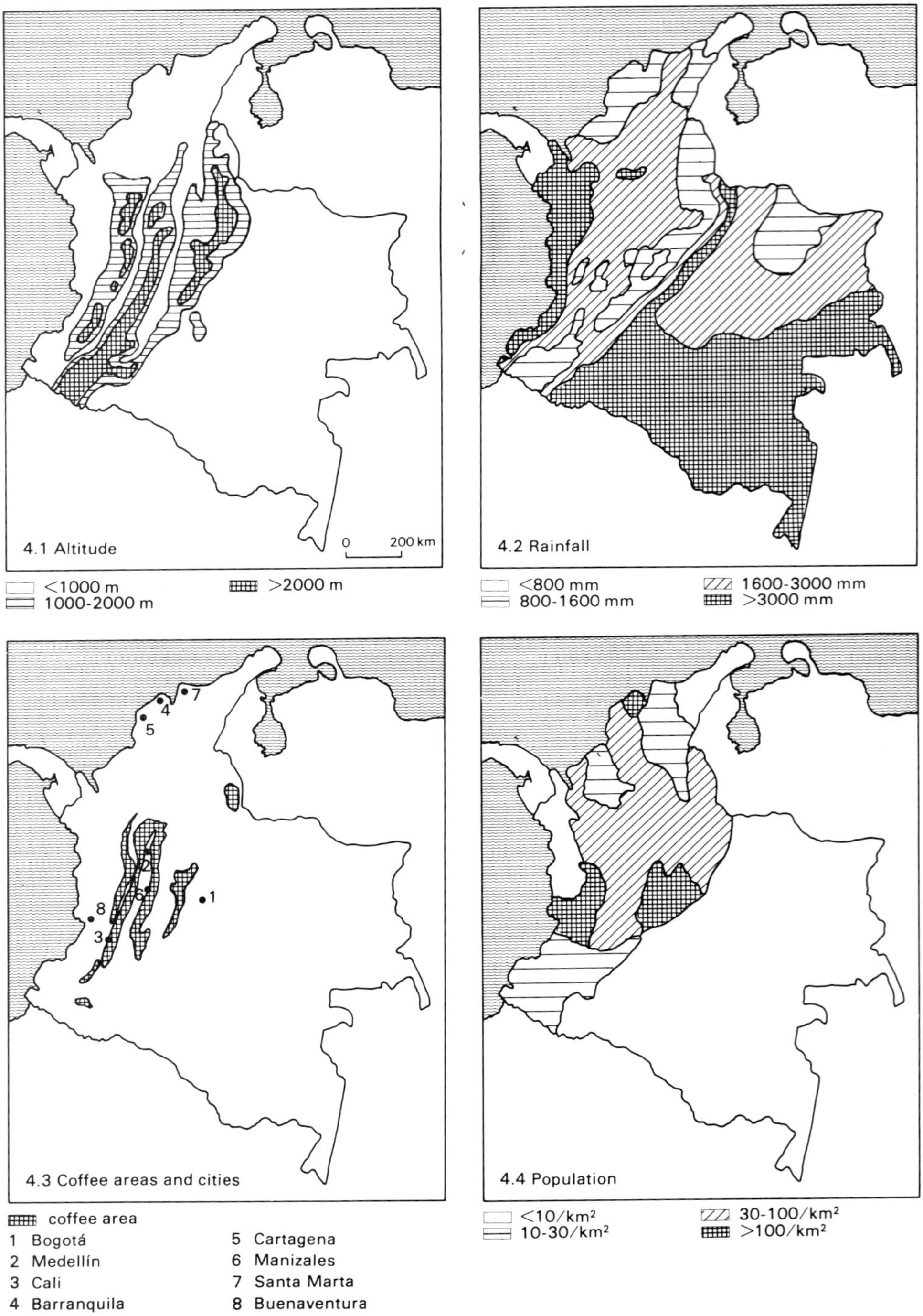

Fig. 4.1-4.4. Altitude (4.1), average annual rainfall (4.2), coffee areas and major cities (4.3) and population density (4.4) of Colombia. After Hulsbosch (1981).

Table 4.1. Average area harvested, yield and national production of major crops over 3-year periods.

Crops	Area (1000 ha)		Yield (kg/ha)		Production (1000 t)	
	1969/1971	1979/1981	1969/1971	1979/1981	1969/1971	1979/1981
Food crops						
rice	260	424	2911	4350	756	1843
maize	684	620	1251	1379	856	854
cassava	185	212	7476	9783	1380	2070
Oil crops						
oil palm[1]	13	26	2590	3530	33	90
soya beans	58	64	1954	2016	114	130
Beverages						
cocoa	45	67	430	535	20	36
coffee	817	1070	591	700	483	748
Other crops						
cotton	241	186	1462	1594	352	297
sorghum	64	219	2407	2219	153	488
sugar cane[2]	241	291	4840	7460	1167	2170
bananas & plantains	.	.	.	.	2407	3403

1. Including palm-oil and palm kernels.
2. Production of sugar centrifugal and non-centrifugal.
Source: FAO (1971-1983)

and the Amazon Area in the east. The Central Highlands represent only 30 % of the total area, but contain 80 % of the population and most of the economic activities. The Caribbean Coastal Area comes second in population and economic activities; in the Magdalena Delta are the major harbour towns. The other regions are largely uninhabitated; the Pacific and Amazon Regions consist mainly of dense tropical rain forest and the Orinoco Area of savanna plains (llanos).

Although completely within the tropics, Colombia contains many climates, from semideserts in the north, through humid tropical climates ('tierra caliente') in the eastern and western lowlands, to temperate (tierra templada) and cold climates (tierra fria) in the Central Highlands. About 80 % of the area falls within the 'tierra caliente' with an average annual temperature of at least 24 °C and altitudes up to 1000 m (Figure 4.1). Another 10 % consists of 'tierra templada', with average annual temperatures between 17 and 24 °C and an altitude of 1000-2000 m. At 2000-3000 m, the 'tierra fria' has temperatures between 12 and 17 °C and constitutes about 8 % of the total area. The remaining 2 % is made up of the mountainous areas and mountain peaks (up to 5750 m), with temperatures below and far below 12 °C, and is called 'tierra helada' or 'paramo' (Hulsbosch, 1981).

The average annual rainfall also ranges widely, from less than 400 mm in the extreme north to more than 8000 mm in the Pacific Coastal Area. In the temperate regions in the Central Highlands, where the main coffee areas are situated, the average annual rainfall is between 1000 and 3000 mm (Figure 4.2). By rainfall distribution throughout the year, this area can be divided into three belts: in the northern belt (6-9°N), a four-month dry period occurs between December and April; in the central belt (3-6°N), two dry periods occur around January-February and around July-August; in the southern part (1-3°N), there is again one dry season from June to September (Table 4.2). Duration of sunshine ranges down from 280 h/month in dry periods to 70 h/month in rainy periods in high-altitude zones. The relative humidity in the coffee zones is 70-85 %.

The optimum conditions for arabica coffee (average annual rainfall of 1800 to 2800 mm, average annual temperatures between 19 and 21.5 °C, and minimum and maximum temperatures around 16 and 26 °C) are found at 900-1400 m, 1200-1600 m and 1400-1800 m in the northern, central and southern coffee belts, respectively. Figure 4.3 shows the major coffee areas, which generally fall within the densely populated regions (Figure 4.4).

Most of the soils in the coffee areas are composed of volcanic ash with a pH between 5 and 6 and have an effective depth of at least 50 cm, a medium to high stability and a medium to low fertility. The other soil types, consisting of basalt, gneis and amphibolites are usually low in stability and are susceptible to erosion. Because of the general steepness of the terrain, erosion is a hazard on all soils. Erosion control is essential on steep slopes, from which shade trees should not be

Table 4.2. Distribution of rainfall for different zones.

Station[1]	Latitude (°N)	Altitude (m)	Annual rainfall (mm)	Monthly rainfall (mm)					
				J	F	M	A	M	J
Pueblo B.	10	1000	2049	30	21	59	126	279	230
Medellín	6	1450	1325	41	49	75	147	178	112
Chinchina	5	1310	2530	135	149	207	276	282	202
Bogotá	4	2556	940	51	50	69	100	105	57
Popayan	2	1789	1911	152	143	168	187	140	98
				J	A	S	O	N	D
Pueblo B.				153	226	282	356	233	54
Medellín				105	133	129	160	114	82
Chinchina				153	169	186	315	261	195
Bogotá				47	41	52	145	138	85
Popayan				40	32	89	267	311	284

1. Stations close to the cities and municipalities mentioned, in the following departments: Cesar, Antioquia, Caldas, Cundinamarca and Cauca.
Sources: Schwerdtfeger (1976), Cenicafé (1980)

removed. The topography also defies mechanization and is a major factor in the exclusive use of the land for coffee.

4.3 The coffee sector

4.3.1 History

In the 18th Century, coffee production in the Western Hemisphere was concentrated in the Caribbean Region, where coffee was introduced by the French (Haiti) and the Dutch (Surinam). The climate in those areas was not optimum for arabica coffee, but transport and shipment were relatively easy. Coffee was cultivated with slave labour in zones less suitable for sugar-cane and bananas. It was gradually introduced into Latin America, first in Brazil, Venezuela and Costa Rica.

It was first planted in Colombia in 1732 by Jesuit priests. Coffee growing was among measures stimulated by a law passed in 1821, which prohibited coffee imports. From the Venezuelan Andes, it entered the bordering provinces of Colombia; from Santander, it spread to regions whose agriculture was based on slaves. The coffee was marketed through Maracaibo or along the colonial trade routes (the Magdalena River) through Santa Marta and Cartagena. The major exports had originally been tobacco, quinine and indigo; cocoa and cotton too were already major crops. In the densely populated Department of Antioquia (originally the centre of gold-mining) and surrounding areas, merchant families had arisen, particularly through the tobacco trade. When tobacco declined in the 1870s, the Antioquia merchants, as well as the Bogota and later the Medellîn merchants began to participate actively in what was then considered 'coffee speculation'. Coffee was an attractive crop since it did not require heavy investment, the technology was easy, the climate and soils suitable, and labour was available.

After 1870, coffee production expanded rapidly over a large area in the Andes, so that railways were built, in turn stimulating coffee production and other economic activities even further. The railways expanded from about 600 km in 1898 to 3300 km in 1940. More than 70 % of what they carried was coffee. On the return trip, they carried imported goods for industrial development (Reveiz, 1980). In 1900, production in Santander still accounted for 60 % of Colombia's coffee but it stagnated, partly because of less suitable ecological conditions, border problems and currency problems. In Santander, coffee production was based on sharecropping on large haciendas and on small family farms (a pattern that emerged after the disappearance of slave haciendas), whereas in Cundinamarca and Antioquia the merchants played a major role in coffee development and soon became large-scale coffee growers and exporters. The size of their estates ranged from about 200 to 2000 ha, one being even 6400 ha.

Cundinamarca and Tolima regions were characterized by a considerable cultural and ethnic difference of landowners ('whites') from day workers and peasant labourers ('indios'). The Arrendamiento System[1] of labour contract was practised, whereas the labourers in Antioquia had an Agregado[2] Contract. Under both types of contract, the tenants obtained a piece of land to be used for food crops. In return, they had to work on the coffee plantations (haciendas). Whereas the minor harvest (mitaca) could be picked by these workers alone, the main harvest required seasonal labourers ('cosecheros', often women), who were usually paid on a piece rate. Whereas landowners in Antioquia had a stable diversified income, those in Cundinamarca and Tolima relied completely on their earnings from coffee, reinvested little and appeared vulnerable in periods of low coffee prices. Their attempts to reduce wages caused social unrest. The first crisis, brought about by a decline in coffee prices, occurred during the War of a Thousand Days (1899-1902) and doubt arose about the future of the coffee business. But in the 1920s, coffee prices increased and production expanded rapidly. Between 1920 and 1930, the area and exports of coffee tripled. However conflicts arose between landlords and tenants, who claimed better working conditions and the right to plant coffee on their own plots. Proposals were then made for land reform with the Land Law (Ley de Tierras). But when this law was finally accepted (1936), it had been so compromised that it was ineffective for many tenants, who lost their contracts, instead of acquiring ownership rights over their land (Meertens, 1979).

The fall in coffee prices in 1930 was a major setback for the Colombian economy, but the prices soon recovered, thanks to measures undertaken by Brazil to decrease its exports and destroy part of its stocks. The next blow was caused by World War II. From 1940 to 1945, the country was urged into a quota agreement. After that, the influence of Federacafé (established in 1927) increased, particularly through the establishment of a national coffee fund (1940) and a national warehousing system.

There was again a coffee boom after the war. In 1954, coffee constituted 83 % of exports. This was during the Civil War ('La Violencia': 1948-1964), which caused migration of landless people to urban areas and decreased prices of land. It also brought about a new law on land rental and an organization for land reform (Incora). The coffee boom had stimulated agro-industrialization, and development mainly in the lowlands of commercial agriculture with new crops (rice, cotton, sugar-cane, sorghum, soya). By this gradual diversification, the share in coffee in exports dropped to only 50 % in 1973. Thereafter, new high-yielding technologies and fast-rising prices on the world market (1976-1978) restored dependence on coffee as the sole export earner of importance.

1. Arrendamiento: system, whereby the worker (arrendatario) receives a small plot of land in return for his obligatory labour on the coffee plantation.
2. Agregado: literally 'attached to hacienda'; worker on hacienda, allowed to cultivate a patch of land as part of his wages.

4.3.2 Recent developments

Coffee production in the 1970s has been greatly influenced by the further introduction, on a large scale, of the Caturra variety. This was accompanied by a more intensive use of inputs and resulted in substantial increases in yields (Table 4.3). Instead of causing a decrease in the national coffee area, the coffee boom during the period 1975/78 reversed the trend and even caused new areas to be brought into cultivation for coffee. Whereas investments by farmers in former periods of high prices resulted in expansion of their coffee area, those in the 1970s generally resulted in a switch towards modern technology. Partly through quick establishment by Caturra, production and exports increased in two to three years, allowing Colombia to benefit substantially from the high prices on the world market in the late 1970s. However after 1980/81, exports were restricted and stocks accumulated rapidly, being equivalent to one year's production in 1983.

Major areas for coffee are the departments of Antioquia, (former) Caldas, Cundinamarca, Tolima and Valle (Table 4.4). Average yields were low in 1970, particularly in the northern and southern departments, but have sharply increased since then with intensification of coffee cultivation.

The considerable differences between the departments in yields can be largely attributed to the extent to which Caturra has been introduced (Table 4.5). The departments, such as Antioquia, 'former' Caldas and Huila, with a large part of their coffee area under modern production systems have the highest average yields. Ecological factors do not differ much between the major coffee departments. Microclimatic conditions and soil conditions vary, however, even within the departments.

Table 4.3. Area planted, national production and average yield of green coffee (1955-1981).

Crop year	Area (1000 ha)	Yield (kg/ha)	Production (1000 t)
1955/56	777	474	368
1970/71	1070	429	459
1971/72	1074	402	432
1972/73	1077	501	539
1973/74	1079	407	468
1974/75	1089	441	540
1975/76	1104	472	480
1976/77	1111	514	558
1977/78	1133	591	648
1978/79	1164	655	720
1979/80	1110*	664	762
1980/81	1010	762	770
1981/82	1010*	792	800*

Sources: Ministerio de Agricultura (1981), Federacafé (1982)

Table 4.4. Area planted and average yield of green coffee by departments in three years.

Department	Area (1000 ha)			Yield (kg/ha)		
	1955	1970	1980	1955	1970	1980
Antioquia	115	155	148	518	465	932
Tolima	116	136	127	483	408	717
Valle	94	127	107	532	488	709
Caldas[1]	187	88	100	627	713	953
Risaralda[1]	.	62	66	627	713	883
Quindío[1]	.	63	62	627	713	1095
Cundinamarca	91	102	89	314	405	647
Cauca	24	79	66	394	296	447
Santander	25	63	56	343	357	660
Huila	45	48	53	344	363	723
Norte de Santander	29	46	44	362	269	594
Other departments	51	101	92	249	282	486
Total/Average	777	1070	1010	474	429	762

1. The Department 'Caldas' was in the early sixties split into Caldas, Risaralda and Quindio.
Sources: ECLA/FAO (1958), Enciclopedia del desarrollo Colombiano (1974), Reportes Ltda. (1982)

Table 4.5. Ecological factors and production system of coffee by department (1980).

Department	Ecological factors			Share of respective productionsysteem (%)		
	lati-tude (°N)	av. alti-tude (m)	av. annual rainfall (mm)	traditional	modern	
					under shade	without shade
Antioquia	6	1450	2100	33	37	30
Tolima	4	1350	2000	73	11	17
Valle	4	1500	1400	76	11	13
Caldas	5	1450	2200	62	12	26
Risaralda	5	1450	2400	66	7	27
Quindío	4	1450	2300	53	20	27
Cundinamarca	4-5	1400	1800	77	14	9
Cauca	2-3	1600	2000	86	9	5
Santander	6-7	1400	1800	72	23	5
Huila	2-3	1450	1300	65	12	23
Norte de Santander	7-8	1300	1700	69	22	8
Avarage of total		1450	2000	66	17	17
Estimated average yield of green coffee (kg/ha)				430	1000	1870

Sources: Federacafé (1982), Cenicafé (1982/83)

4.3.3 Organization

In Colombia, coffee is now produced on more than 300 000 farms, ranging in area from less than 1 ha to more than 100 ha. Most farms are run by families and are less than 4 ha, but more than 50 % of the national coffee production comes from medium-sized and large farms (above 20 ha).

The coffee cherries are usually processed into parchment coffee on the farms, most of which possess pulping and drying facilities. The dried parchment is taken to the nearest village or town, where it is sold either at one of the 500 purchasing points of Federacafé or to traders who operate independently or for exporters. Independent traders can decide whether to sell to the Federation or to private exporters. The Federation buys parchment coffee through its agents at a fixed minimum price, as long as it is of a certain quality. After collection, the parchment is bulked and transported to depots or mills. It is cured just before export.

The National Federation of Coffee Growers (Federacafé) was established in 1927, as a private non-profit association of coffee producers. Through contracts with the Government, it gained influence on national policy in the coffee sector. The Federation is responsible for the management of the National Coffee Fund, for provision of technical assistance to growers, for the control of domestic and export marketing, and for advice on the setting of certain rates of taxation and prices for the coffee industry. Its supreme authority is the National Congress of Coffee Growers. Technical assistance, education and social services are provided by Departmental Committees. A major organization within Federacafé is Almacafé, responsible for the extensive coffee warehousing system in Colombia. The Federation also sponsors and administers the National Coffee Research Centre (Cénicafé) and is involved in the provision of credit to farmers and in the organization of producer-cooperatives. At central level, Federacafé is structured, under a General Manager, into two 'subgerencies' and four 'gerencias' (Directorates), responsible respectively for administration; commercialization; planning and institutional coordination; and development and diversification (Prodesarrollo). The first three are principally responsible for the coffee sector, whereas the fourth is in charge of the coffee diversification programme.

The National Coffee Fund, which obtains its considerable revenues from taxes on coffee, was established in 1940, originally to finance the surplus stock. The functions of the Fund have been enhanced greatly over time. It has also become a major investor in infrastructural facilities and activities related to the coffee sector, and it supports the coffee institutions.

The Government itself is neither involved in the production, nor in the processing and marketing of coffee, but close relations with Federacafé are established through 'contratos de servicios', which are periodically renewed and by which Federacafé takes over the responsibility for coffee production and marketing. Federacafé is remunerated for these services by the publicly owned Coffee Fund, through special taxes and levies. The Government is permanently repre-

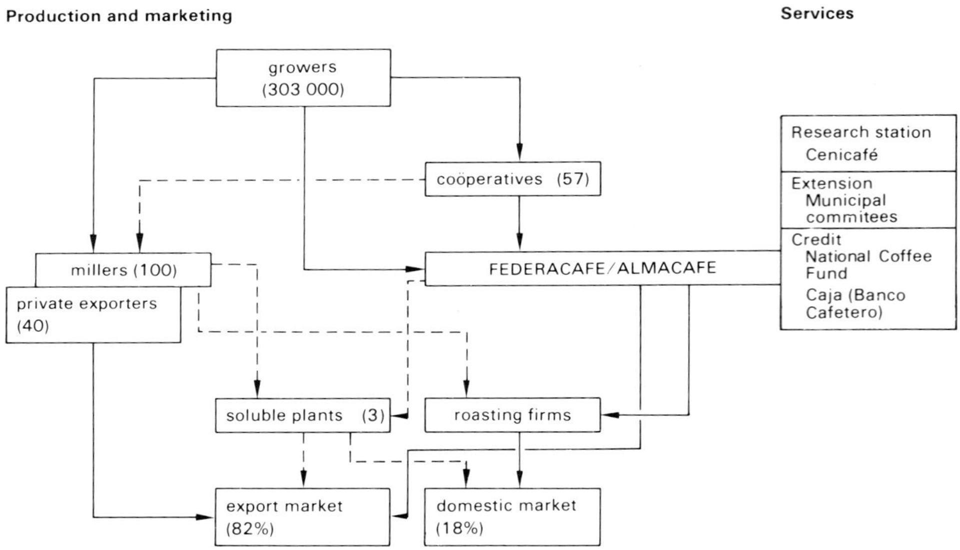

Fig. 4.5. Participants in the coffee sector and main marketing channels in Colombia in 1982. Solid line, main channel; broken line, secondary channel.

sented in the National Coffee Growers Committee, which is charged with the general supervision of the activities of Federacafé.

Figure 4.5 shows the participants in the coffee sector and the main marketing channels for coffee in Colombia (Section 4.6.1).

4.4 Production

4.4.1 Production units and farming systems

The coffee zone in Colombia comprises all areas along the three mountain chains of the Andes, with an altitude 1000-2000 m. According to that definition, the zone covers about 4.8 million hectares or 4.2 % of the national area. About 75 % of the coffee is produced under agro-ecologically optimum conditions at altitude 1300-1800 m. In those areas, nearly all farms, or a total of 303 000 cultivate some coffee: on average 3.5 ha out of an area of almost 15 ha (against a national average of 26 ha). In 1955/56, the number of farms in the coffee zone was only 235 000, with an average area of 20 ha and a slightly smaller average coffee area of 3.3 ha.

But both farm size and area under coffee vary considerably (Table 4.6). Nearly half the farms are less than 4 ha (minifundia). Despite their large involvement in coffee cultivation, they contribute only about 12 % to national production. Large farms (above 20 ha), with only 10-20 % of their land under coffee, are more diversified but produce half the national coffee crop. The average yields in 1970 did not

Table 4.6. Proportion (%) of number of farms, area under coffee and production by farm sizes.

Farm size (ha)	Share of total number of farms	Total area	Share of total coffee area	Share of total coffee production	Share of residing population[2]
less than 1	13	79	2	2	7
1-4	33	57	12	11	33
4-8	18	43	12	12	15
8-20	19	35	23	23	19
20-100	15	22	36	36	19
more than 100	2	11	15	16	7
Total average	100	24[1]	100	100	100

1. 1 070 000 out of 4 460 000 ha.
2. On average 1.8 person per hectare under coffee or 6.3 per farm.
Source: Federacafé (1974)

vary much with farm size (about 400 kg/ha for small and 500 kg/ha for large farms), but the amount of parchment coffee produced per family labourer ranged from about 70 kg for small farms to 340 kg for medium-sized farms and 580 kg for large farms. By region, the smaller farms are found in Cundinamarca, Cauca and the former department Caldas (average farm size 10 ha). Nevertheless about 45 % of the average farm size is under coffee in Caldas 'Viéjo', against 39 % in Valle, 27 % in Cundinamarca and only 18 % in Antioquia.

Now, more than 85 % of the coffee farms are operated by the owner, covering more than 70 % of the total area under coffee, whereas about 13 % is run by managers or tenants (28 % of total coffee area). In 1950, 14 % of the farms (20 % of the area) were still sharecroppers, but this practice is no longer widespread. The other major crops in the coffee zone were in 1970 sugar-cane, plantain, cassava and maize, but the major alternative land use was pasture with more than 45 % of the total area. Of minor importance were cocoa and beans (Table 4.7). Whereas other permanent crops and annual crops respectively 6 and 3 % of the total land use respectively, and featured on both small and large farms in similar proportions, pasture was predominantly found on the farms of 50 ha and above. The importance of the crops and pasture varied by region, but pasture usually constituted between 35-50 % and sugar-cane between 3-7 % of land use. About 24 % of the land in the coffee zone was not used and also not suitable for agriculture.

Much attention is paid to diversification, buth this can hardly be achieved with the crops now grown in the area. Crops like cocoa and sugar-cane are not suitable for the high-altitude mountainous coffee zone and the annual crops often cause much erosion on the steep slopes in those areas. More could be expected of the further introduction of such crops as pineapple, citrus, avocado and other tree crops (spices and fast-growing forest species) and, in areas close to markets, to

Table 4.7. Land use in coffee zone (1970).

Land use	Total area (1000 ha)	Proportion of land use (%)	Number of farms (× 1000)
Coffee	1070	22	303
Sugar-cane	201	4	57
Plantains[1]	53	1	26[2]
Cassava	.	.	27[2]
Maize	158	3	19[2]
Beans	.	.	3
Cocoa	40	1	3
Pasture	2169	45	217
Other use or waste	1085	24	.
Total	4776	100	320

1. Plantains are sometimes interplanted in coffee plantations.
2. To some extent mixed cropping of plantains, cassava and maize.
Source: Federacafé (1974)

dairying, small livestock and other fruit-trees. In densely populated zones, tea would be possible as well but, given the present wage rates, it is not likely that Colombia could compete with the Asian tea-producing countries. Coffee is also interplanted with some other crops and particularly with plaintains, which are applied as nurse crops (shade for young trees) and are sometimes retained when the trees have become mature. Under traditional coffee cultivation, little use is made of chemical fertilizers and labour inputs are confined to sporadic weeding, pruning and spraying.

4.4.2 Production stages and production costs

4.4.2.1 Establishment

As the terrain where coffee is planted is generally mountainous, land preparation for new plantations or replanting is usually performed with handtools. It has been estimated that in the period 1970-1978 more than 60 % of the new coffee plantations consisted of replantings and less than 40 % of plantings on land not previously used for coffee. For replanting, use was traditionally made of young coffee trees that sprouted spontaneously and relatively few had their own nurseries. Now, more use is made of small nurseries with 8 000-32 000 seedlings, where the new short and early maturing varieties are multiplied. Large nurseries are generally avoided to prevent spread of diseases.

Research carried out by Cénicafé on the costs of production of seedlings, through the germination of seed in seedboxes and the raising of seedlings on seedbeds, showed that the costs per coffee plant were of the order of 0.90 pesos in

1975 (Lopez-Alzate et al., 1976). Calculations based on those data, show that the costs per plant in 1983 were about 4 pesos for medium-sized nurseries of about 30 000 seedlings and about 5 pesos for farmer's own small nurseries with 6500 seedlings. Labour constituted about 60 % of these costs (Annex 4.1).

Apart from land clearing, major cost elements in the establishment of modern coffee plantations without shade, are construction or soil conservation measures (e.g. basins, interception drains) and the removal of trees (that had served as shade trees). The lack of shade also requires higher labour inputs for weeding during establishment and the productive period. For traditional plantations, plant density is usually 1000-1500 coffee trees per hectare, whereas (without shade) a planting density of 5500 trees per hectare is now aimed at for modern Caturra plantations. This great difference in plant densities also influences the costs of es-

Table 4.8. Establishment costs (years 1-12) per hectare for traditional and modern (without shade) coffee plantations (1983).

Cost elements	Traditional			Modern				
	1	2	3/4	establishment			stumping	
				1	2	3	11	12
Labour inputs (man-days)								
clearing & lining out	40	–	–	40	–	–	–	–
soil conservation	–	–	–	30	10	10	10	10
removal shade	–	–	–	30	–	–	–	–
(re)planting	20	2		70	6	–	–	–
weeding	–	70	70	–	110	80	80	80
shade regulation	–	20	20	–	–	–	–	–
fertilization	–	–	–	–	35	35	35	35
spraying	–	–	–	–	7	–	7	–
stumping	–	–	–	–	–	–	35	–
total	60	92	90	170	168	125	167	125
Subtotal (1000 pesos)	19	29	29	65	64	48	63	48
Material inputs								
number of plants	1500	80	–	5500	450	–	–	–
value of plants (1000 pesos)	7	1	–	22	2	–	–	–
fertilizers: (kg compound)	–	–	–	–	1125	1375	1250	1375
(kg urea)	–	–	–	–	125	125	125	125
value fertilizers (1000 pesos)	–	–	–	–	25	30	28	30
chemicals (1000 pesos)	–	–	–	–	2	–	2	–
Subtotal (1000 pesos)	7	1	–	22	29	30	30	30
Costs per year (1000 pesos)	26	30	29	87	93	78	93	78
Total establishment costs (1000 pesos)		114 (yrs 1-4)			258 (yrs 1-3)			
Stumping costs (1000 pesos)					171 (yrs 11-12)			

Source: Fedesarrollo (1979)

tablishment, and is accentuated by the high rate of fertilizers, as required from the first year onwards for modern intensive cultivation. Under this system, the coffee trees should be stumped at an age of 10 years, resulting in a two-year unproductive period. For traditional plantations, which only start bearing in the fourth year, the establishment costs consist mainly of landclearing, planting, weeding and regulation of shade. Shade trees are usually *Unga* spp. or plantains, the last of which are also common as temporary shade, during establishment. Table 4.8 gives establishment costs for both traditional and modern coffee plantations.

4.4.2.2 Productive period

A major distinction can be made, as indicated earlier, between three coffee production systems, or levels of technology: the traditional production system, with small material inputs and corresponding low yields; a modern intensive production system without shade, with a high planting density and a high rate of fertilizers, and a modern intermediate production system, in which some shade is retained, intermediate planting densities are adopted and relatively modest rates of fertilizers are used. The relationship between coffee prices and prices of fertilizers determines, whether a modern system with or without shade is preferred. Between these two modern production systems are a range of options, in which some shade is kept at a planting density of about 4200 trees per hectare, which is allowed to develop in the periods of unfavourable price ratio of coffee to fertilizers, and which is reduced when coffee prices become more attractive.

For both traditional and modern coffee farms, the annual costs of production consist largely of labour inputs, particularly for weeding and even more for harvesting. In most coffee areas in Colombia, the main harvesting season is mid-September to December, and a secondary minor harvest in May or June. These are peak periods of demand for labour. During the main harvest, 50-100 kg of cherries are picked per man-day. Climatic conditions are favourable for selective picking. In Caldas ('Viéjo') a considerable part of the demand for picking labour used to be met by migrant workers, who combined coffee picking in Caldas with cotton picking in other areas (Cauca Valley). Many small farmers are underemployed in the periods between harvests, and this constitutes a major consideration for the diversification programmes, which try to introduce other crops and other sources of income on such farms.

Whereas on modern plantations without shade compound fertilizers are usually applied (usually 18-6-18-2 at 1500 kg/ha, in split dressings), sometimes supplemented with urea, small traditional farmers hardly supply any chemical fertilizers, but they often use coffee pulp as manure. On modern plantations with shade, compound fertilizers are used at about 500 kg/ha or less. Pruning is not regularly done on small farms, and is not important on modern farms either, where stumping is carried out after 6-7 years of production. Although several pests and diseases occur in the coffee zones, spraying with insecticides and fungicides is a mi-

Table 4.9. Production costs of green coffee (per hectare and per kilogram) for production systems: traditional, modern under shade (tecnificado al sombra) and modern without shade (tecnificado al sol).

	Traditional	Modern under shade	Modern without shade
Yield (kg/ha)	430	1000	1875
Plant density[1] (plants/ha)	1500	3500	5000
Cyclus (productive years)	4-22	4-10 & 14-20	3-9 & 12-18
Wage rate (pesos/man-day)	320	360	380[2]
Annuity of establishment costs (1000 pesos) (10 %, 20 years)	12.2	26.0	34.7
Annual costs			
Labour inputs (man-days)			
weeding	28	48	60
fertilizing	2	12	22
disease control	3	3	2
pruning & shade control	10	10	4
soil conservation	1	2	5
harvesting (main & sec.)	46	78	120
processing	8	14	24
transport	2	2	3
administration	1	2	4
other activities	3	3	6
total	104	174	250
Subtotal of costs (1000 pesos)	33.3	62.6	95.0
Material inputs (1000 pesos)			
fertilizers & manure	2.0	10.0	29.6
insecticides & funcigides	0.7	1.4	2.2
processing	1.5	3.0	4.5
Subtotal	4.2	14.4	36.3
Overhead costs (1000 pesos)			
depreciation & interest equipment	2.0	4.2	6.5
interest on loans	1.0	3.2	7.8
administrative costs	1.3	2.6	4.2
water & electricity	1.2	2.7	3.6
social security payments	3.4	10.0	17.8
other (fees)	0.4	0.5	0.7
Subtotal	9.3	23.2	40.6
Total costs per hectare (1000 pesos)	59.0	126.2	206.6
Costs per kilogram (pesos)	137	126	110
(US$)	1.81	1.66	1.45

1. According to 1980-census respectively 1800, 3800 and 4600 plants per hectare; while in several other studies densities of 1000 and 5500 are adopted for traditional and modern (without shade).
2. The basic wage is 300 pesos, but with fringe benefits 380 pesos.
Sources: Cenicafé (personal communication)

nor activity, if at all. In some areas, ant control is required. Since the advent of processing of coffee cherries into parchment, coffee in Colombia has generally been processed on the farm, and therefore the costs of them are also included in annual costs (Table 4.9). Annual costs of production, excluding establishment, are 92-109 pesos per kilogram of green coffee and, including establishment costs, 110-137 pesos per kilogram.

4.4.3 Yields and returns

Over the period 1950 to 1970, average yields fluctuated around 500 kg of green coffee per hectare for adult plantations, and average yields for all coffee land were just below 500 kg/ha and were related to the extent of new plantings. With the ageing of trees in traditional plantations, yields are likely to fall and, despite replanting within traditional plantations, yields will probably remain at about 430 kg/ha. The yields of Caturra variety in modern plantations will depend on the age of the trees, timely stumping and on the application of fertilizers, and on other inputs and cultural practices. The mean yield of all productive coffee land on which modern cultivation practices are applied, is of the order of 1500 kg of green coffee per hectare, and the increasing share in modern plantings has boosted national average yields to around 750 kg/ha in 1980. Under the most intensive coffee cultivation systems, without shade and with heavy dressing, average yields can be obtained of around 2000 kg/ha, from the 3rd to the 9th year and from the 12th to the 18th year (after stumping). But in the absence of stumping, with lower inputs of fertilizers and with shade, yields will remain in the order of 1000 kg/ha.

The price a producer receives for his parchment coffee depends first on whether he sells to agents of exporters or to the Federation and its cooperatives. Section 4.6.1 will show that there is usually a price differential, making it attractive to sell to one or the other. But the price also depends on quality. In fact, only one grade is accepted, referred to as 'café tipo Federación' for which 1280 pesos was paid per arroba or 102.40 pesos per kilogram in May 1983 (arroba is a measure containing an equivalent of 12.5 kg of parchment coffee). The grade is determined by colour, humidity, number of damaged beans and abnormal beans. A maximum of 5 % of low-grade beans is accepted under the standard grade. Since on average, 15 to 20 % of the beans supplied by the farmer consists of 'pasillas' or 'ripios' (lower-quality beans), the farmer should either sort the beans to reach the standard or sell the coffee at a discounted price (as 'café corriente'). Pasilla coffee is mainly used for the domestic market and in May 1983 fetched only about 100 pesos per 'arroba'. On average, farmers received only about 1150 pesos per 'arroba' for their coffee or 92 pesos per kilogram. It is likely that modern progressive farmers received slightly more (about 95 pesos per kilogram) and traditional farmers, with lower-quality processing equipment, slightly less for their coffee (about 90 pesos per kilogram).

This would imply that in 1983 coffee farmers could not generally cover integral costs (including establishment), but farmers using modern technolgy (without shade) covered their annual costs, excluding establishment costs. If it is assumed that farmers who apply traditional cultivation methods use their own family-labour for about 60 % of coffee activities, the average return for labour would have been about 190 pesos per man-day in 1983, or almost half the wage paid to hired labourers (excluding establishment costs). Total employment in coffee production can be roughly estimated at 600 000 man-years or two man-years per farm. Total gross earnings from coffee for the average farm in the coffee areas is about 250 000 pesos (1983), which probably constitutes about a third to 40 % of gross earnings.

4.5 Processing

The processing of coffee is largely decentralized: the initial (wet) processing is mostly on the farm, and the curing (hulling and grading) is undertaken by the exporters. Almost all coffee producers possess at least the minimum equipment for processing of coffee up to the parchment stage. This equipment usually consists of a hand-operated or diesel-operated depulping machine, one or two tanks in which coffee is fermented and washed, and a drying yard or drying trays, and some other equipment. During the main harvest, depulping is done a few times a week, followed by fermentation (one and a half days) and washing. Sun-drying in a yard might take up to a fortnight. Because of the rainfall distribution and the spread of picking activities over several months, much labour is required, particularly from women and children, to protect the coffee from rain during drying. Therefore use is also made of movable trays (heldas). Calculations by Cénicafé showed that the costs of establishing drying tables for 225 kg of parchment were of the order of 430 pesos in 1974 (about 1300 pesos in 1983). Modern farms use mechanical steam-heated dryers. The costs of on-farm processing have been already mention in Section 4.4.

Several studies by Cénicafé on the technical and economic feasibility of central processing factories showed that such factories would offer economies of scale. However they should be planned to operate near capacity, because of the relatively high fixed costs involved (depreciation and interest would constitute about half of the processing cost, excluding raw material).

Hulling and grading takes place in about 130 factories, many operated by exporters. Federacafé (Almacafé) operates 9 factories and authorizes about 85 private factories to process its coffee. The margin for the hulling operation was around 130 pesos per 70-kg bag of green coffee in 1983. The capacity of the hulling installations varies from about 250 to 1200 bags per day. But the limiting factor in this processing stage is formed by the classification activity, which is usually carried out manually (by women). One person can classify about 95 kg of coffee per hour (1.2-1.5 bag). The total installed hulling and grading capacity is about 20 million bags per year. Total employment is estimated to be around 2 million man-days or almost 10 000 man-years.

Roasting is predominantly for locally consumed coffee, but a certain amount of semiroasted and roasted coffee is exported. The roasters buy the coffee mainly from the Federation at a subsidized price. There are a few processing factories in the country for soluble coffee, which is sold on both the domestic and the export market (Nescafé and Colcafé). The Fondo Nacional de Café owns a freeze-dried coffee plant, in Chinchina. This plant has a capacity of 2000 t of soluble coffee per year and provides employment for about 300 labourers. An expansion, undertaken in 1983, will double the capacity and enhance employment by 50 workers. Because of contractual obligations of the Government with the other plants, this plant has to export all its produce, for which it fetched a price of about US$8-10 per kilogram in 1983. Statistics show that the plant has operated far below capacity in recent years.

4.6 Marketing and pricing

4.6.1 Marketing channels and marketing functions

Depending on the quality of their coffee and the prices offered, the farmers sell it either at one of the 500 buying points of the Federation and its (57) cooperatives, or to traders and agents of exporters. The traders bulk the parchment and transport it to one of 40 exporters, including the Federation. The exporters store the coffee and have it milled at one of about 100 factories just before export. Because of taxation procedures, the lower qualities of coffee generally accrue to the Federation, which sells them to roasters (at a subsidized rate) for sale to domestic consumers. This coffee is further classified as consumo, pasilla and ripio. But most of the parchment coffee bought by the Federation is for export and samples are sent to the 'Departamento de Calidades' for analysis and classification (respectively as Klaus, Europa, Escandinavia or Americano). This classification is on the basis of bean size and number of defects; the largest beans (screen 16.5) and highest quality (Klaus) are meant for the West German market. When an export order comes in, another sample is taken for final classification, after which the coffee is hulled and sold to the international dealers or roasting companies. The Federation also exercises control over the quality of the coffee bought by private agents for export by the private export firms.

One of the major responsibilities of the Federation in recent years is storing excess production (above the export quota), which accumulated rapidly in recent years. Its subsidiary firm Almacafé runs about 150 warehouses; of these 70 are owned by the organization, and have a total area of 250 000 m^2 and a capacity of 11 million bags. The others with an area of 100 000 m^2 and a capacity of 4 million bags are rented. There are privately managed warehouses for storage of coffee, with a total capacity of about 5 million bags. A new development is that of inflatable tents, four of which are now in use, with an area of 8000 m^2. In April 1982 about 17 million bags of parchment coffee (equivalent to about 12.7 million bags of green coffee) were stored and about a million 50-kg bags of fertilizer. With the high turnover of stores, storage costs in 1982/83 were only about 0.27 pesos per kg per month. Almacafé has a total of about 1800 employees (including 270 for hulling operations).

Now, at least 90 % of transport from the warehouses to the port of shipment is by road, against 10 % by rail. From its inception, the Federation has been anxious about transport problems, and played a major role in the establishment of shipping companies for inland transport (Navenal in 1944) and for ocean freight ('La Flota Mercante Gran Colombiana'). The latter company was originally owned by Colombia, Venezuela and Ecuador together, but Venezuela pulled out in 1953.

4.6.2 Pricing and taxation

The private exporter has to surrender the foreign exchange received from the export of coffee to the Central Bank within twenty days of registration for export. After deduction of the ad valorem tax, the Bank exchanges these proceeds for currency-exchange certificates, which can be converted to pesos immediately at a discount of between 6 and 15 % or after 120 days at their full value. To ensure that the amount of foreign exchange corresponds to the actual earnings of coffee that enter the country, a certain amount of foreign exchange has to be surrendered to the Bank for each 70-kg bag of green coffee exported. This is called the 'reintegro'. The reintegro is about equal to the foreign-exchange earnings per bag of coffee exported and is adjusted with varying prices on the world market. But the reintegro sometimes lags behind prices, allowing exporters to accumulate foreign currency dwing price rizes or forcing exporters to purchase foreign exchange dwing falling prices. In April 1983 the reintegro was US$191.

Before obtaining permission to export, exporters have to pay three different taxes: the ad valorem export tax; the retention tax; and the 'pasilla and ripio' tax. The ad valorem export tax has to be paid to the Central Bank in foreign exchange and amounted to 9 % of the reintegro value in May 1983. Part of this tax is subsequently paid to the treasury (5 % in May 1983) and part to the National Coffee Fund, through which some payments are made to the regional committees of coffee growers. For the payment of the retention tax and the 'pasilla and ripio' tax, exporters have to submit a certain amount of parchment coffee to warehouses of the Federation (Almacafé). The retention tax dates from 1958 and was originally meant to withhold some coffee from the market in view of the Latin American coffee agreement. The rate has often changed and amounted to 40 % of the amount of coffee to be exported, in May 1983. The receipts constitute the major financial resources of the National Coffee Fund. Under the 'pasilla and ripio' tax, exporters have to deliver 5.5 % of their consignment as 'pasilla' and 0.5 % as 'ripio' parchment, for which they are paid a fraction of its value. In Table 4.10 the marketing margins and taxes are given, for four different periods.

Table 4.10. Prices, marketing costs and margins in US$ per kilogram (green coffee equivalents) in four periods.

	July 1975	July 1977	April 1979	April 1983
ICO indicator price for Colombian milds	1.79	5.35	3.05	3.00
Export unit value (f.o.b.)	1.69	5.23	2.97	2.82
(Reintegro)	(1.69)	(5.24)	(2.82)	(2.73)
Cost and margins				
ad valorem tax[1]	0.32	0.88	0.46	0.25
retention tax[2]	0.40	2.42	0.91	0.76
pasilla & ripio tax	0.01	0.02	0.01	0.01
costs of hulling	0.02	0.02	0.02	0.02
costs of transports & storage (incl. margin)	0.10	0.09	0.08	0.11
Total	0.85	3.43	1.48	1.15
Price to growers (at collection point)	0.84	1.80	1.49	1.67
Price to growers as share of f.o.b. price (%)	50	34	50	59

1. Ad valorem tax respectively 19 %, 17 %, 16 % and 9 % of reintegro value.
2. Retention tax: for each kg green coffee to be exported to deliver respectively 35 %, 80 % and 40 % of an equivalent (factor 0.7) amount of parchment coffee.
Source: Federacafé (1980, 1982)

4.6.3 Export and domestic consumption

In the past two decades, about 18 % of national production has been used for domestic consumption. This locally consumed coffee consists mostly of the lower qualities. But to boost consumption, efforts are being made to improve average quality. Consumption per person per year is about 4 kg of green coffee equivalent, which is usual in major coffee-producing countries and not much less than in major importing countries. Table 4.11 shows the distribution of the national production over consumption, exports and stocks.

Of the 40 exporters, the Federation is the major one with about 40 % of total exports in recent years. Only one private exporter ships more than a million bags per year, whereas about 12 companies export between 100 and 500 thousand bags per year, and more than half of the exporters export less than 50 000 bags per year. The best known trade mark of Colombian coffee is MAM, which stands for Medellín, Armenia and Manizales, some major production areas. Nearly 60 % of total coffee exports are shipped from Buenaventura, followed by the two major ports of Santa Maria and Cartagena with 24 % and 16 %, respectively. Major buyers of Colombian coffee are the United States and West Germany, with the Netherlands in the third place. In 1981/82 more than a third of total coffee exports were destined for West Germany (Table 4.12).

With the rapid increase of production in recent years and the international ex-

Table 4.11. National production, domestic consumption, exports and stocks (in 1000 t) (1973-1982).

Crop year	National registered production	Total production	Consumption	Export	Final stock
1973/74	442	468	93	444	150
1974/75	480	540	96	453	141
1975/76	436	480	84	421	116
1976/77	542	558	84	318	272
1977/78	629	648	96	453	371
1978/79	694	720	98	686	307
1979/80	753	762	102	693	274
1980/81	770	770	108	540	396
1981/82*	774	780	110	550	516
1982/83*	760	770	110	560	616

Sources: Ministerio de Agricultura (1981), Federacafé (1982)

Table 4.12. Proportion distribution (%) of coffee exports to major importing countries averaged over two periods.

	1977/78-1980/81	1981/82
United States	27	18
West Germany	28	35
Netherlands	9	8
Sweden	5	5
Japan	4	5
Spain	4	4
others	23	25

Source: Federacafé (1982)

port-quota system in operation to stabilize prices, Colombia is being faced with an enormous overproduction. Part of this finds its way through illegal exports through third countries, but tremendous stocks have to be kept, which now almost equal a year's production (Table 4.11).

4.7 Supporting services

4.7.1 Research and extension services

The Colombian Federation of Coffee Growers (Federacafé) is involved in coffee marketing, in many other activities related to coffee. It sponsors and administers the coffee research station Cénicafé (Centro Nacional de Investigaciones de Café) in Chinchina, which carries out research on ecological, agronomical and economical aspects of coffee production, and also covers coffee processing. It established a centre for agricultural education, the Manuel Mejia Foundation, and operates a chemical laboratory. Federacafé is also responsible for extension services to coffee farmers. Extension officers work at municipal level. The municipal committees are the smallest territorial units of the Federation and can make requests to the Departmental Committees (14 in total) for rural development activities to be financed by the Coffee Fund. Allocation is based on the share in total coffee production.

4.7.2 Provision of credit

Since the establishment of the 'Caja de Crédito Agrario, Industrial y Minero' in 1931, financial assistance to farmers in the coffee areas has gradually become a major supporting service. The Caja captures not less than about 40 % of the national institutional savings and is supported financially by the Federation and the Fondo Nacional del Café. The volume of agricultural loans outstanding in 1980 amounted to over 36 000 million pesos, a third of which originated from the Caja and about 60 % from the Fondo Financiero Agropecuario (FFAP), which channels part of these funds through the Caja. To complement the activities of the 'Caja' the Federation established the Banco Cafétero in 1953, which operates with funds from the Fondo Nacional del Café. The total volume of loans outstanding in 1977 exceeded 11 thousand million pesos, about a third of which was directly used for promotion of coffee production. About a thousand million pesos was channelled in that year through a special revolving fund, benefiting about 20 000 small and medium-sized coffee farmers. In 1983, with inflation reaching about 17 % per year, farmers could obtain loans to a maximum of 80 000 pesos for renovation of one hectare of pure-stand coffee, at an interest rate of 23 %.

Considerable funds are now allocated to the Diversification Programme (Prodesarrollo). Other credit and savings institutions include the Concasa (Corporacion Cafétera de Ahorro y Vivienda), agricultural insurance companies and some financial corporations.

4.7.3 Input supply and cooperative development

The major material inputs in coffee production are fertilizers and agro-chemicals. Until 1978, the country produced all of the phosphorus and a large part of the nitrogen and compound fertilizers used in the country. But in 1978, total sales went up from about half a million tonnes in previous years to three-quarters of a million tonnes. These fertilizers are manufactured and distributed by a small number of companies, the major ones being Abocol and Monomeros. Abocol belongs to the twenty largest companies in the country. The Caja Agraria is also involved in production and sale of fertilizers and also operates five seed-production stations for food crops. Federacafé has stimulated the development of coffee cooperatives, which combine marketing, credit and input supply functions with other socio-economic activities. In 1982, there were 57 cooperatives with about 550 service centres and about 130 000 members with a total paid capital of almost 600 million pesos, and sales of coffee and consumer goods (including inputs) of 43 and 10 thousand million pesos, respectively.

4.7.4 Diversification programme

The diversification programme was created in 1963, to provide financial and technical assistance to foster diversification in coffee areas and to promote a more balanced economic development. The diversification fund was originally a private organization, but in 1968 it was placed under the Ministry of Agriculture to fulfil the diversification obligations of Colombia as a member of ICO. As from 1975, the programme was extended to all coffee departments and became dependent on the Federation. Four five-year diversification plans have been implemented since then and a fifth is under preparation and is being presented to the World Bank for co-finance. Prodesarrollo has studied soil capability systematically in the coffee districts to obtain a technical basis for diversification in the various ecoclimatic areas. Unfortunately few agricultural activities combine the characteristics of coffee in ecological requirements, employment, foreign-exchange earnings, 'value-to-weight' ratio in marketing and erosion prevention. Therefore the total package includes many different agricultural, forestry, livestock and processing activities.

4.8 Summary and economic parameters

Colombia is the second largest producer of coffee in the world and in 1972-1982 accounted for 14 % of world production. The climatic and soil conditions in the central highland, the most densily populated area of the country, are very suitable for arabica coffee and rainfall patterns are such that selective picking and wet processing can be spread over several months, thereby making effective use of labour.

Consequently coffee is the most important crop in the country, both in terms of area cultivated and value of production. Coffee contributes to about half of total exports, so the national economy depends very much on its coffee production and on prices of coffee on the world market. Suitable land for coffee has become scarce, and to reduce the use of marginal land, causing erosion, emphasis is now on more intensive cultivation systems using the compact variety 'Caturra'. The country produces almost exclusively the high quality Colombian Mild arabica coffee. Wet processing is mainly done on the farm, and hulling is undertaken by exporters.

A major organization in the coffee sector is Federacafé (Federation of Coffee Growers), which is involved in regulatory activities, marketing and export of coffee, research and extension, diversification programmes and finance of infrastruc-

Economic parameters of the coffee sector in Colombia (1982).

Resources used in coffee sector	
Land under coffee	1.1 million ha
as share arable & permanent crop land	18 %
Labour involved	700 000 man-years
in production	630 000 man-years
in marketing, processing & services	70 000 man-years
as share of agricultural work force	5 %
Material inputs for coffee production	12 000 million pesos*
Income from coffee sector	
Value of coffee production	
(volume × export unit value)	160 000 million pesos
as share agricultural production	25 %
as share of GDP	7 %
Value of coffee exports (f.o.b.)	1 577 US$ million
as share of agricultural exports	73 %
as share of exports	51 %
Share of total production value, obtained by	
Producers	55 %
Processors, traders & exporters	8 %
Government and parastatal organizations	37 %
Other parameters	
Production Arabica (Colombian milds)	100 %
Yield (green coffee)	770 kg/ha
Production of labour involved (green coffee)	4.0 kg/man-day
Share of coffee produced for export	82 %
Ratio producer price (per kg green coffee):	
daily wage rate (pesos/man-day)	0.26

Sources: FAO (1971-1983), IMF (1983), World Bank (1984) and own estimates

tural projects. The organization and its regional committees receive funds from the National Coffee Fund, which obtains its financial resources largely from various taxes on coffee. With intensification of coffee cultivation, production will probably continue to increase soon, which will further complicate the situation of large stocks and illegal exports, given the international export quota. Because of the increasing scarcity of suitable land, there is, however, little scope of further increase in area under coffee, and continuous efforts should be made towards diversification.

References

Cenicafé, 1980. Avances Tecnicos, March 1980. Federacafé, Chinchina.

Cenicafé, 1982/83. 40 Años de investigación en Cenicafé; Volumen 1 – Suelos and Volumen 2 – Clima. Federacafé, Chinchina.

ECLA/FAO, 1958. Coffee in Latin America; I Colombia and El Salvador. FAO. Rome.

Enciclopedia del Desarrollo Colombiano, 1974. El café. Collection Los Fundadores, Volume III, Bogotá.

FAO, 1971-1983. Production Yearbook. FAO, Rome.

Federacafé, 1974. Data from 1970 coffee census. Bogotá.

Federacafé, 1982. Cuadragesimo primero congreso nacional de cafeteros, Informe del Gerente General (and annexes). Federacafé, Bogotá.

Fedesarrollo, 1979. Economia cafetera Colombiana. Fondo Cultural cafetero. Bogotá.

Hulsbosch, F.N.M., 1981. Landendocumentatie 1981, no. 5, Colombia. KIT, Amsterdam.

IMF, 1983. International Financial Statistics. IMF, Washington.

Lopez-Alzate, R. & Gallo-Cardona, A., 1976. Avances Tecnicos, no. 48 & no. 51. Cenicafé, Federacafé, Chinchina.

Meertens, D., 1979. Jonkers en boeren; de strijd om het land in Colombia. CEDLA, Amsterdam.

Ministerio de Agricultura, 1981. Cifras del sector agropecuario, 1980. Bogotá

Reportes Ltda, 1982. Datos de Colombia 1982, annuario estadístico. Bogotá.

Reveiz, E., 1980. Universidad de Los Andes, Facultad de Economia; La Cuestion caféte-ra. CEDE, Bogotá.

Schwerdtfeger, W. 1976. World survey of climatology, Volume 12: Climates of Central and South America. Elsevier, Amsterdam.

World Bank, 1984. World development report. Oxford University Press, Washington.

Currency equivalents and price indices in Colombia (1955-1983).

	Exchange rate (pesos per US$)	Consumer price index (1980 = 100)
1955	2.5	3
1960	6.6	5
1965	10.5	9
1970	18.4	15
1975	30.9	34
1976	34.7	40
1977	36.8	54
1978	39.1	63
1979	42.5	79
1980	47.3	100
1981	54.5	127
1982	64.1	159
1983 (May)	76.4	191

Source: IMF (1983)

Annex 4.1. Costs (in pesos) of production of seedlings for different sizes of seedboxes and nurseries (1983).

Stage I. Germination of seed in seedboxes (duration: two months)

	Number of germinated seeds		
	35 000	12 000	7 500
Area required (m^2)	9	3.5	2
Material inputs			
seed (and in kg)	900 (9)	350 (3.5)	200 (2)
boxes, materials, insecticides	810	520	410
Labour inputs (and in man-days)	2 870 (7.5)	1 260 (3.3)	960 (2.5)
Interest on capital (18 %)	820	380	280
Other costs, losses etc. (10 %)	540	250	190
Total	5 940	2 760	2 040
Costs per seedling	0.20	0.26	0.31

Stage II. Raising of seedlings (duration: 5-6 months)

	Number of seedlings		
	30 000	10 500	6 500
Area required (m^2)	900	350	220
Material inputs			
polyethylene bags	12 000	4 200	2 600
other	4 880	2 310	1 670
Labour inputs (and in man-days)	61 120 (150)	26 360 (69)	16 810 (44)
Interest on capital (18 %)	14 040	5 920	3 790
Other costs, losses, etc. (10 %)	9 200	3 880	2 490
Total	101 240	42 670	27 360
Costs per seedling raised[1]	3.75	4.51	4.68
Total costs per plant[1]	3.97	4.81	5.03

1. Only 90% of initial number of seedlings.
Source: Lopez-Alzate & Gallo-Cardona (1976)

5 Coffee in Costa Rica

5.1 Introduction

With 2.3 million inhabitants and a total land area of 51 350 km², Costa Rica is not densely populated. But large differences in population density exist between the 7 provinces and the 5 planning regions, with about 150 inhabitants per square kilometre in the Central Region against 15 only in the northern and atlantic regions (Huetar-regions/Limon Province).

The GNP per person reached US$1430 in 1982. The share in the major sectors: agriculture, industry and services in GDP was respectively 25 %, 27 % and 48 %; and in total employment 29 %, 23 % and 48 % (World Bank, 1984). The contribution of coffee production to GDP was about 5 %.

About 2 million hectares, or 40 % of the total land area is agricultural and pasture land, distributed as follows: 283 000 ha (6 %) arable land, 216 000 ha (4 %) under permanent crops and 1536 000 ha (30 %) pasture. About a million ha consists of forest and bushland, one fifth of which is under exploitation. More agriculturally potential land could be brought under cultivation, particularly in the northern and atlantic zones.

In Table 5.1 the major crops in Costa Rica are shown. The contribution of agricultural products to the total value of exports (f.o.b.) fluctuated in the period 1974/1980 between 63 % and 73 %, and the share in coffee exports in total exports ranged from 20 % to 40 % during that period. Coffee is the major export product, followed by bananas. It concerns for almost 100 % arabica coffee, although a little robusta coffee is produced in the lowlands.

With an average annual production of 109 000 t in the period 1979/81, Costa Rica contributed about 2 % of the world production of coffee and ranked 11th as world producer just after India and Uganda and before Cameroon.

5.2 Ecology

Costa Rica is situated within the tropics between 8° and 11° North of the Equator, and forms part of the Central American isthmus with on the east and west-side, 200-300 kilometres apart, respectively the Atlantic and Pacific Oceans. Despite its size and location, Costa Rica contains many different climates and types of vegetation, largely due to the influence of the three mountain chains. These chains contain several volcanoes, the highest of which is the Irazu (3500 m) in the

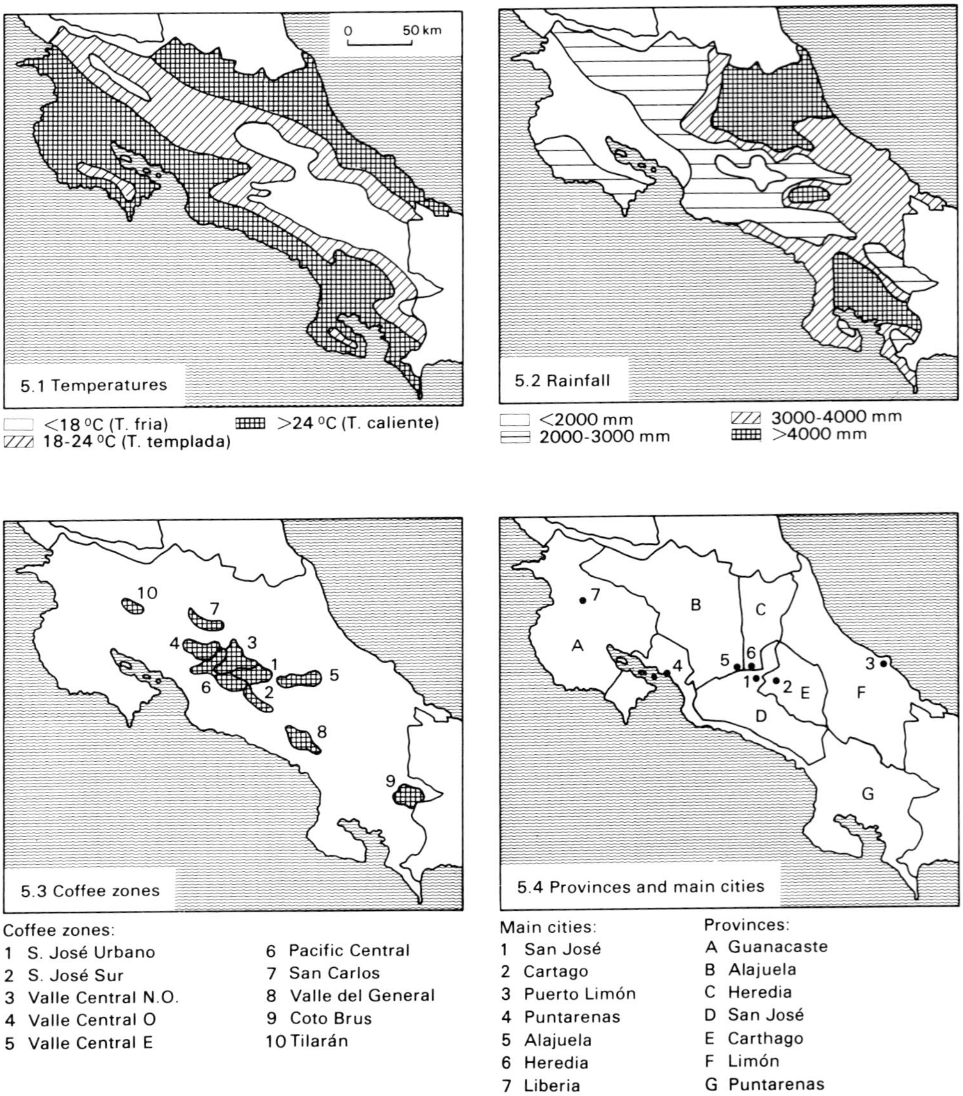

Fig. 5.1-5.4. Average annual rainfall (5.1), annual mean temperatures (5.2), coffee zones (5.3) and provinces and major cities (5.4) of Costa Rica. After KIT (1974) and CONICIT (1982).

Table 5.1. Average area harvested, yield and national production of major crops over 3-year periods.

Crops	Area (1000 ha)		Yield (kg/ha)		Production (1000 t)	
	1969/1971	1979/1981	1969/1971	1979/1981	1969/1971	1979/1981
Food crops						
maize	50	49	1123	1609	56	79
rice	44	73	2019	2734	88	199
sorghum	6	24	1639	2085	11	50
Oil crops						
coconuts	.	.	.	.	17	25
oil palm	.	.	.	.	16	33
Beverages						
cocoa	30	18	179	331	5	6
coffee	95	83	870	1315	82	109
Other crops						
bananas*	28	25	39900	44200	1119	1105
cotton	1	9	1637	1452	2	13
sugar-cane[1]	36	48	4917	4500	177	216
tobacco	2	2	923	954	2	2

1. Production of sugar centrifugal and non-centrifugal.
Source: FAO (1971-1983)

Central Highlands. The largest mountain range, the Cordillera de Talamance, in the south east has peaks over 4000 m. Except for a few larger ones, the valleys in between the mountains are relatively small, and communication has always been a major problem.

On the lower areas along the coast and in the north average annual temperatures are above 22 °C. These areas are referred to as Tierra caliente (Figure 5.1). In the north-east, along the Atlantic Coast and in the south these temperatures are combined with a high and well-distributed annual rainfall (3500 mm and above), which makes these zones suitable for the production of such crops as oil palm, cocoa and bananas. But the coastal zone in the north-west with the 'Peninsula de Nicoya', and the 'Valle del Tempisque' is hot and dry, with an annual rainfall of 1500 mm and less (Figure 5.2). In the mountainous areas in the interior of the country average annual temperatures are generally 10-22 °C (Tierra templada and Tierra fria) and in the areas above 3300 m in the south-east even less than 10 °C (Tierra helada).

Arabica coffee is grown mainly in the 'Tierra templada' zones, on an altitude between 600 and 1500 m, generally accompanied by average annual temperatures between 16 and 20 °C. At higher altitudes, the danger of frost exists, and at lower altitudes the flowering is hampered. The optimum rainfall is 1500-2500 mm, and although coffee is also grown in zones with higher rainfall (Turrialba), the high humidity in such areas causes a higher incidence of fungi-diseases and more problems in weed control. The distribution of the mean annual rainfall (in mm) for San José, at an altitude of 1120 m, is as follows:

Total	Month:	J	F	M	A	M	J	J	A	S	O	N	D
1944		8	5	10	37	244	284	230	233	342	333	172	46

Whereas the total rainfall is abundant in this region, the dry season is long, and therefore supplementary irrigation is often beneficial (Section 5.4.2). The daily mean temperature in San José is 20.4 °C, and the extreme maximum and minimum temperatures are 33.2 and 9.4 °C, respectively.

Coffee in Costa Rica is classified by type of bean, depending on hardness, acidity and aroma, and these qualities vary according to climatological conditions found in the respective coffee areas in the country (Table 5.2).

Among the most common soil types in Costa Rica are the lateritic soils (Ferralsols), which are chemically poor because of leaching, the for agriculture even less suitable Lithosols in the mountainous areas, and the Plenosols in the coastal and other lowland areas, which are mainly suitable for rice. The most fertile land is found on the alluvial and vulcanic soil types. Alluvial soils are found both along the Atlantic and the Pacific coast, and are mainly used for banana and oil palm plantations. Volcanic soils are found mainly around the Cordillera Central. These soils are very appropriate for coffee production because of the organic material and the texture which facilitates drainage as well as cultivation. Coffee is also grown on lateritic soils, but these contain little humus and are often too heavy. The combination of soils of vulcanic origin in a temperate highland climate with about 9 humid months in most production areas, constitute very favourable conditions for coffee growing.

Table 5.2. Classification of coffee beans and coffee areas and average annual climatic data.

Type of coffee bean	Altitude (m)	Rainfall (mm)	Number of days with rainfall	Temperature (°C)	Sunshine (h/yr)
Strictly Hard (SHB)	1200-1600	2500	155	19	2150
Good Hard (GHB)	1000-1200	2250	160	21.5	2200
Hard Bean (HB)	800-1200	2500	158	22	2100
Medium Hard (MHB)	600-1100	3500	185	22	1800
High Grown Atlantic (HGA)	900-1200	2750	210	20.5	1700
Medium Grown Atlantic (MGA)	600- 800	2900	245	22	1750
Low Grown Atlantic (LGA)	300- 600	4000	245	24.5	1550
Pacific (P)	300-1000	2250	145	24	.

Source: Oficina del Café (1976)

5.3 The coffee sector

5.3.1 History

When Costa Rica became independent in 1821, (until 1848 first as part of the Central American Federation), the country had only some 50 000 inhabitants, most of which from Spanish descent. This population was concentrated in the Meseta Central (Central Valley). They cultivated food crops (maize, wheat), kept pastures for animal production, and grew some cashcrops (cocoa, sugar-cane and tobacco). But through regulations and taxes, the Colonial Government (Audiencia de Guatemala) had discouraged the production of sugar-cane and tobacco. Costa Rica lacked the precious resources the Spanish had found in other Central American countries such as Guatemala and El Salvador: minerals and a large population of Indians to be employed on the plantations. Those countries were important producers of indigo and cochineal up till 1860-1870, when the world market for those products declined. By that time coffee, cultivation of which was not discouraged by taxes, etc. had already become a major exportcrop in Costa Rica.

Since independence the national development has in fact been markedly influenced by the two agricultural products, bananas and coffee, that dominated exports for more than a century (1840-present). Whereas bananas were mainly grown in the coastal areas on large estates by foreign companies, that employed negroes from the Caribbean, coffee was the major cashcrop for the farmers in the relatively densely populated central valley.

Major problems in the early development of coffee production were the lack of contacts with commercial firms and traders in Europe for both coffee marketing and provision of credit, the lack of appropriate processing facilities to obtain good export quality, the lack of labourers (particularly) during harvesting and the bad connections of the highlands with the harbours on the Pacific and Atlantic coasts (Hall, 1982). Traders gradually established contacts with importers in England, which remained the main market for Costa Rican coffee until World War II. During and also after the war much coffee was sold to the United States.

With ample land available Costa Rica lacked the work force required for the labour-intensive production and processing of coffee. Contrary to Brazil and the Caribbean islands slaves had not been imported and by law of 1862 it was prohibited to import and employ negroes and Chinese. As a consequence wages of labourers were high. It was therefore important to produce high export-quality coffee and Costa Rica was among the first countries to adopt the method of wet processing. This originally took place on the farm, but was later transferred to processing factories. These were owned by some of the larger coffee producers and their families, but also by traders and exporters, among which some foreigners.

The construction of main roads and railways from the Central Highlands to the harbour towns, has been extremely important for the development of the coffee industry. Although it meant for ships to pass Cape Horn, the easiest connection was that to the Pacific Coast, where climatic conditions were also better. In 1846 a road was built from San José to Puntarenas, later (in 1910) followed by a railway line along that road. The least cumbersome connection to the Atlanctic coast would have been along the river San Juan, on the border with Nicaragua, but it was decided to construct a railway line between San José and Limon, which was started in 1871 and completed in 1890.

Once the infrastructure was established and the necessary conditions fulfilled the coffee production increased rapidly, until 1935 mainly in the central valley, and afterwards also in other areas. The area under coffee (usually expressed in manzanas; 1 manzana = 0.69 ha) increased from 18 000 ha in 1890 to 47 000 ha in 1935 and reached about 90 000 ha in 1980. Although cocoa was originally (before 1840) a more important crop than coffee and could be grown in many areas, its development has been hampered by several factors. In the first place because of the scarce population in the ecologically suitable tropical lowland areas. Secondly large estates for several reasons preferred to produce bananas and oil palm. And in some of the cocoa areas the rural infrastructure remained inadequate: cocoa farmers paid taxes for the construction of a main road, but that was never built. Finally, most owners of cocoa plantations did not live on their farms, and left the management to tenants.

5.3.2 *Recent developments*

During the period 1930 to 1950 the production and area under coffee remained stagnant, because of the world depression and World War II. From 1950 onwards the total area planted increased rapidly, and yields doubled. In the 1960s cultivation techniques improved and the traditional Typica variety was gradually replaced by higher yielding varieties, particularly the Hibrido Tico. This was followed in the 1970s by the introduction of the small Caturra trees. This resulted in higher planting densities per hectare (about 3400 trees per hectare in 1980 against 1600 in 1955). With the increased use of fertilizers, yields of green coffee went up from about 1000 kg/ha in 1970 to 1300 kg/ha in 1980. But the recent incidence of leaf-rust will have a profound effect on the coffee production soon.

Table 5.3 shows the development of the area planted, the national production and average yields over the period 1970-1981. The proportional distribution of coffee production by type was in 1980/81 as follows: Strictly Hard Bean: 37 %; Hard Bean: 22 %; Medium Hard Bean: 16 %; Good Hard Bean: 11 %; Atlantics: 7 %; Pacific: 1 %.

Ten main coffee zones can be distinguished, which together cover about 94 % of the total area under coffee. These areas are indicated on Figure 5.3 and their relative importance is shown in Table 5.4. The table illustrates how new coffee areas, such as the 'Valle del General', 'Coto Brus' and 'San Carlos', gradually de-

Table 5.3. Area planted, national production and average yield (1970-1981).

Crop year	Area (1000 ha)	Production (1000 t)	Yield (kg/ha)
1970/71	75.5	73.2	950
1971/72	76.7	89.4	1140
1972/73	77.9	78.8	1000
1973/74	80.4	95.7	1170
1974/75	83.4	84.3	1000
1975/76	81.8	80.2	960
1976/77	81.8	78.1	950
1977/78	81.0	91.3	1150
1978/79	81.0	104.9	1300
1979/80	81.8	91.1	1110
1980/81	82.5	121.7	1470
1981/82	82.5	103.9	1250

Source: Oficina del Café (1982)

Table 5.4. Coffee zones and main types of coffee produced.

Zone	Proportion of total area (%)			Type[1]
	1935	1950	1973	
S. Jose Urbano	27	26	11	SHB
S. Jose Sur	8	10	7	SHB
Valle Central N.O.	24	24	20	HB/SHB
Valle Central C.	9	9	13	SHB
Valle Central E.	23	20	16	SHB/M&HGA
Pacifico Central	8	9	13	HB
San Carlos	–	0	3	LGA
Valle del General	1	2	12	MHB
Coto Brus	–	0	5	SHB/MHB
Tilarán	–	–	10	P/HB/MGA

1. See Table 5.2 for description of the types.
Source: Aguilar et al. (1982)

veloped outside the Central Valley. Some of these new coffee areas, now covering about 15-20 % of the total area, are because of the altitude and climate less suitable for arabica coffee. Because of the modern techniques applied in these new areas, this is not always reflected in the yields, but it is brought out by the average quality.

5.3.3 *Organization*

Coffee is in Costa Rica produced by about 33 000 farmers and all coffee is processed in some 100 processing factories (beneficios) spread over the coffee areas. These factories are generally involved in all processing stages: from depulping and washing to hulling, grading and bagging. Producers sell their coffee cherries through collectors to these factories, which after processing sell the green coffee to exporters, or take care of the export themselves. The factories also sell some coffee through the Oficina del Café to local roasting firms which supply the local market.

The Oficina del Café exercises extensive control over the production and marketing of coffee in Costa Rica. The organization was established by law in 1948 to control the relations between producers, processing firms and exporters. Many regulations exist in the coffee trade, one of the major of which is found under law no. 2762 of 1961, prescribing that coffee producers should establish a contract of sale with a factory (beneficiador), which is obliged to pay him the minimum price fixed every year by the 'Junta de Liquidaciones'. Every year the Oficina del Café

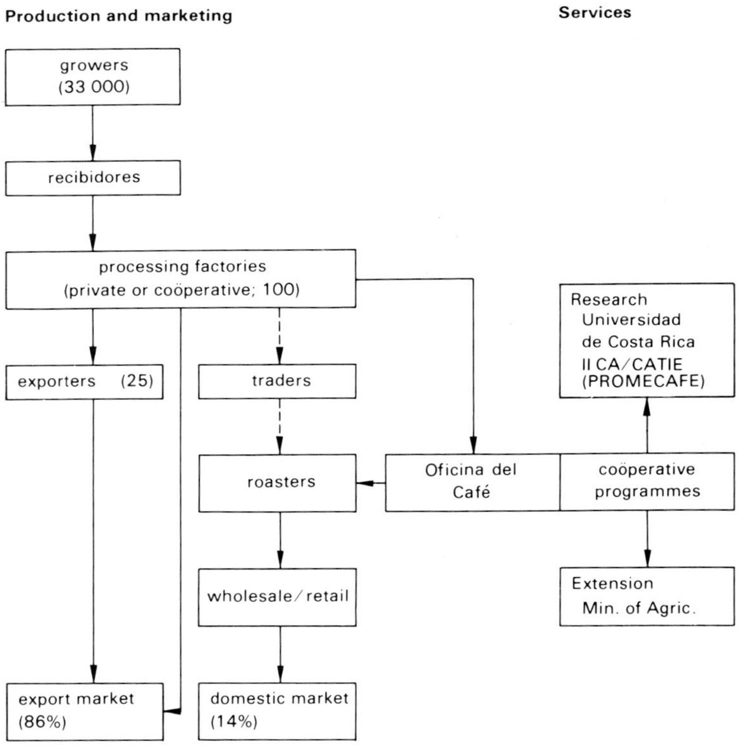

Fig. 5.5. Participants in the coffee sector and main marketing channels in Costa Rica (1982). Solid line, main channel; broken line, secondary channel.

carries out some studies on the costs of production and on the costs incurred by processing factories, to be able to fix the margins for exporters and 'beneficiadores' and, by taking into account the prices on the world market and the taxes, also the minimum price to producers.

The Ministry of Agriculture (MAG), is responsible for the extension services, and since 1982 holds the presidency in the Junta Directiva Oficafé. Agricultural credit is supplied by the National Banking system which is also involved in several investment programmes in the coffee sector. Research is carried out by the 'Centro de Investigaciones en Café' (Cicafé), the 'Facultad de Agronomia' of the University of Costa Rica, and by IICA and CATIE, specifically through their programme 'Promecafé'. Coffee producers and 'beneficiadores' are organized in the 'Camara Nacional de Cafétaleros', and in cooperatives, which are part of the 'Federación de Cooperativas RL'. Every year the respective participants in the coffee sector meet in the 'Congreso Nacional de Cafétaleros'. The Government is not directly involved in the production and processing of coffee, but plays through the respective governmental organizations, mentioned above, a major regulatory role in the marketing and pricing of coffee.

5.4 Production

5.4.1 Production units

The 1973 census showed a total of 81 600 farms in the country, together covering 3.1 million hectares or an average of 38 ha per farm. This also includes land not used and unsuitable for cultivation and/or pasture. The distribution is quite uneven with more than 50 % consisting of small farms (less than 10 ha) owning about 5 % of the land and 3 % made up of the large farms (more than 200 ha) which own more than 50 %. This distribution is largely due to the high concentration of population and farming activities in the Central Highlands and to the importance of the estate and livestock sectors, respectively in the coastal area and in the north-west part of the country. The important crops in the estate sector are bananas, oil palm, rice and sugar-cane.

In the Central Highlands about 70 % of the farms are smaller than 10 ha, whereas the highest production is obtained from farms between 10-50 ha, refered to as 'familiar' (Table 5.5). Because of historical factors relatively little land in Costa Rica is rented, although farmers not always possess the necessary title deeds on the land they operate. In the early period of coffee growing in Costa Rica, farmers could be granted ownership on land, after having grown coffee for a period of more than 5 years.

Coffee is grown on about 33 000 farms (40 % of the total), with on average about 2.6 ha under coffee. The small farms (less than 2 ha) have coffee plots of only 0.5 ha on average, whereas the average size of coffee plantations on large farms (of about 100 ha) is 50 ha. Whereas the highest productivity per hectare in

Table 5.5. Proportional distribution (%) of the production of main crops by farm size (1973).

Farm size (ha)	Maize	Coffee	Cocoa	Sugar-cane	Rice	Bananas
less than 10 ha (small)	26	30	14	11	6	1
10-50 ha (familiar)	39	31	35	16	17	3
50-200 ha (medium)	23	22	27	18	18	8
more than 200 ha (large)	12	17	24	55	59	88

Source: Dirección General de Estadística y Censos (1974)

Table 5.6. Proportional distribution (%) by farm size of number of farms, area cultivated, production and average yields of green coffee (1973).

Farm size (ha)	Number of farms	Area (ha)	Production (t)	Yield (kg/ha)
All farms	32353	83407	85000	1020
less than 2 ha	37	8	6	800
2- 10 ha	35	27	24	910
10- 50 ha	21	31	31	1010
50- 200 ha	6	18	22	1240
200-1000 ha	1	12	14	1170
more than 1000 ha	(0.2)	4	3	830

Source: Aguilar et al. (1982)

1955 was found on the largest farms, in 1973 the medium-sized farms (50-200 ha) had the highest yields. Table 5.6 shows the development of number of farms, areas and coffee production and the distribution of them by farm size.

5.4.2 *Production systems*

Coffee is predominantly produced by small farmers, sometimes as a monoculture, but more often in combination with other crops, in particular with the food crops maize and beans (IICA-Protaal, 1982). The producer prices for maize and beans have been kept low for a long time resulting in a decrease of area harvested and relatively few technological improvements. Other farming activities often combined with coffee are dairying and cultivation of sugar-cane, which employs labour when coffee does not.

Since the late 1960s important technological changes took place in the coffee production. These were made possible through the institutional arrangements and the research and extension activities that had been initiated, and because of the fact that coffee growing was considered more profitable and less risky than other crops. These technological improvements can be divided into:

– Biological factors (better genetic material).
– Increasing use of chemicals (fertilizers, insecticides).
– Agronomical measures (higher planting densities, soilconservation and the more effective use of shade).
– Mechanization (including the use of irrigation).

Traditionally coffee in Costa Rica was of the Typica variety and was grown under shade (Inga and Erythrina-trees). Relatively few farmers used fertilizers and even then only at reduced rates. Gradually high-yielding varieties were introduced, mainly of the Bourbon type, such as Villa Sarchi, Hibrido Tico, Geisha and from Brazil: Mundo Novo and Caturra. In 1968 the extent of the Typica variety was reduced to 47 % of the area, and the Hibrido Tico and Caturra variety covered 37 % and 9 % of the area. Thanks to the absence of the major coffee diseases (rust and berry disease), high planting densities were then not risky, and combined with the further introduction of smaller plants, such as Caturra, offered many advantages (e.g. weed control, higher production per hectare). Caturra varieties indeed covered 33 % of the total coffee area in 1979, against 28 % for Typica and 23 % for Hibrido Tico. Whereas with large plants (Typica) planting densities are usually around 1500 trees per hectare, with Caturra this can be augmented to 7000 trees per hectare. But since the establishment costs are high, many farmers have gradually interplanted more Caturra trees, and only large farms have 100 % Caturra plantations. The average planting density increased from 1525 pl/ha in 1935 to 3365 plants per hectare in 1979.

This new intensive production system had to be accompanied by higher levels of labour and material inputs (fertilizers, spraying, pruning, etc.). In 1979 not less than 76 % of farmers (with 85 % of coffee area) applied fertilizers at an average rate of 880 kg/ha. In 1980 about 50 % of farmers sprayed fungicides and about 70 % applied herbicides to reduce labour inputs for weed control.

Under the intensive production system with high fertilization, shade becomes less important. But only between 10-20 % farmers have eliminated the shade trees, and most farmers have adopted a policy, in which they heavily cut back shade trees in years of high coffee prices (when it pays to increase the rate of fertilization), and, in which they leave shade trees spread themselves in periods of low prices. With the introduction of other varieties and other planting systems the pruning methods changed as well, and some of the these now allow the use of motorized pruning equipment.

Because of the sloping nature of the terrain a mechanization oriented approach of coffee production is not feasible in Costa Rica. To improve the distribution of the water supply and to reduce erosion, irrigation (both sprinkler and gravity) is applied in some areas. Although the annual rainfall at higher altitudes is normally sufficient, the rainy season is often delayed just when the flowers are in bud in April. Irrigation at that moment brings the blossom out, and through the timing of irrigation the picking can be spread out to some extent.

Picking is highly seasonal (3-4 months) but is nevertheless done selectively to produce pulpable ripe cherries, to be sold to the private pulperies. The Low Grown Atlantic and the Medium Hard Bean types of coffee are harvested mainly in the period September-November, whereas the Strictly Hard Bean type is generally not harvested before the period December to February.

5.4.3 Production cost

The economic analysis of coffee production in Costa Rica, carried out every year by the 'Departamento de Estudios Técnicos y Diversificacion' of the 'Oficina del Café' presents a good picture of production costs for coffee in the respective coffee zones and for five different farm size groups[1]. In these studies a distinction is made between four main cost categories: labour costs for cultivation activities; costs of chemicals; costs of harvesting and transport; and other costs, including depreciation and interest on equipment, administration costs and interest on the capital invested.

In Table 5.7 these costs are presented under slightly different headings: labour inputs, material inputs and overhead costs, whereas establishment costs are indicated separately. Total annual costs per hectare were in 1981/82 on average about 38 000 colones, equivalent to about US$1500, and average costs per kilogram green coffee were about 35 colones (about US$1.40). Whereas the costs per hectare for the small farms were only about 60 % of those for large farms, the costs per kg of green coffee produced by small farms were 150 % of those produced by large farms, since the latter had average yields of about 1700 kg/ha against 650 kg/ha obtained by small farms.

In Table 5.8 the average production costs are given for the respective coffee production zones. Whereas the lowest costs per hectare were obtained in some of the new, and in fact less suitable coffee areas such as San Carlos, Valle El General and Guanacaste, the lowest costs per kilogram green coffee were in 1981/82 found in Coto Brus and in the Valle Central Oriental. High cost coffee areas are San José Urbano, Valle Central Noroeste (both because of their high land value) and Valle Central Occidental, San José Sur and Pacifico Central. The increasing share in fertilizers and other material inputs in total costs in the 1970s is best illustrated by comparing the proportional distribution of the three respective cost categories: labour inputs, material inputs and overhead costs, which changed from 63 : 19 : 18 in 1973/74 through 49 : 25 : 26 in 1976/77 to 42 : 39 : 19 in 1981/82. The relatively high overhead costs in 1976/77 were due to the price boom in that period, which gave way to relatively large investments.

1. The strata are in fact not defined by farm size but by the volume of coffee cherries delivered: less than 10, 10-20, 20-80, 80-200 and more than 200 m^3. The classes are locally expressed in 'double hectolitres' (1 dhl=0.2 m^3).

Table 5.7. Production costs of green coffee (per hectare and per kilogram) for five farm size groups (1981/82).

	I	II	III	IV	V	Average
Size of plantation (ha)	1.2	2.3	3.6	7.8	33.4	3.8
Yield[1] (kg/ha)	650	970	1420	1760	1670	1080
Plant density (plants/ha)	2970	3290	3570	3990	4260	3360
Wage rate (colones/man-day)	67	72	72	77	85	72
Annuity of establishment costs (1000 colones) (10 %, 20 years)	10.3	10.7	13.7	12.8	19.0	11.9
Annual costs						
Labour inputs[2] (man-days)						
weeding & application of herbicides	16	18	17	17	9	17
fertilizing	6	6	7	7	6	6
disease control	2	5	7	6	3	4
pruning	13	14	15	13	10	14
regulation shade	7	8	8	6	6	7
harvesting	53	74	115	132	113	85
transport	7	10	13	17	14	11
other[3]	9	9	10	9	10	9
total	113	144	192	207	171	153
Subtotal of costs (1000 colones)	7.6	10.4	13.7	15.8	14.6	11.0
Participation family labour (%)	60	50	30	10	3	40
Material inputs (1000 colones)						
fertilizers (8 ¢/kg)[4]	4.7	6.5	8.4	8.4	8.5	6.8
fungicides (200 ¢/kg)	0.2	0.8	1.7	2.0	0.9	1.0
herbicides (100 ¢/kg)	0.6	0.8	0.9	1.3	1.2	0.8
insecticides (70 ¢/kg)	0.0	0.1	0.2	0.3	0.1	0.1
plants for replanting	1.4	1.5	1.9	1.2	1.6	1.6
Subtotal (1000 colones)	6.9	9.7	13.1	13.2	12.3	10.3
Overhead costs (1000 colones)						
depreciation & interest equipment	1.7	1.7	1.6	1.6	0.8	1.6
interest on loans	0.4	0.5	0.6	0.6	0.5	0.5
administration costs	1.6	1.6	1.8	2.2	2.7	1.8
taxes	0.2	0.1	0.1	0.2	0.2	0.1
other	1.0	0.7	1.1	1.6	1.7	1.0
Subtotal	4.9	4.6	5.3	6.2	5.8	5.0
Total costs per hectare (1000 colones)	29.7	35.4	45.8	48.0	51.6	38.2
Costs per kilogram (colones)	45.60	36.40	32.22	27.24	30.91	35.30
(US$)	1.38	1.10	0.98	0.83	0.94	1.07

1. Usually expressed in double hectoliter cherries, which give about 23 kg green coffee; the yields from the sample were below the national average.
2. 1 man-day = 8 man-hours; labour inputs for harvesting are derived from total harvesting costs divided by average wage rate.
3. Including: maintenance pathes, soilconservation, replanting, etc.
4. ¢ = colones.
Source: Oficina del Café (1983a)

Table 5.8. Costs and returns of green coffee production, by coffee zone (1981/82).

	Average of total	Coffee zone[1]									
		I	II	III	IV	V	VI	VII	VIII	IX	X
Size of plantation (ha)	3.8	4.5	3.3	2.8	8.3	4.6	2.6	2.7	2.7	3.4	1.4
Yield (t/ha)	1.1	0.8	1.3	1.0	1.0	1.2	0.8	0.9	1.7	1.1	0.9
Total gross value (1000 colones/ha)	58	42	70	58	57	62	41	48	91	61	48
Total costs (1000 colones)	32	38	43	38	54	45	28	29	47	33	27
Total net value (1000 colones/ha)	20	4	26	20	3	17	13	10	44	28	21
Value (colones/kg)	54	55	54	56	55	52	55	52	55	53	56
Costs (colones/kg)	35	53	34	37	52	38	37	31	28	29	31

1. Coffee zones: I San José Urbano, II San José Sur, III Pacifico Central, IV Valle Central Noroeste, V Valle Central Occidental, VI San Carlos, VII Valle El General, VIII Coto Brus, IX Valle Central Oriental, X Guanacaste.
Source: Oficina del Café (1983a)

5.4.4 Yields and returns

In the 1970s average yields of coffee increased gradually but given the technological improvements, this increase was not substantial. However the yields were in 1970/71 already at a high level (Tables 5.3 and 5.9). The production costs per kilogram green coffee did show a similar gradual increase. A peak occurred in the period 1976/77, when the price boom stimulated farmers to invest more in their coffee plantations.

The producer prices generally remained well above the production costs for the average producer, but in 1979/80 the margin was small. Table 5.8 shows that through large variations in yields, the returns vary considerably by region. In such high cost areas as San José Urbano and Valle Central Noroeste total costs (including interest on capital invested) were in 1981/82 only just covered. And in 1979/80 the total production costs in those areas were certainly higher than the average producer prices obtained in that year.

5.5 Processing

5.5.1 Processing factories

Coffee processing in Costa Rica is centralized and all processing activities from depulping, fermenting and drying to curing and bagging, take place in about 100

Table 5.9. Yields, producer prices and production costs per kg green coffee (1970/71-1981/82).

Crop year	Yield (kg/ha)	Producer price (colones)		Production costs (colones)			
		current (prices)	1980 constant (prices)	labour, materials, adm.		total[1]	
				current (prices)	1980 constant (prices)	current (prices)	1980 constant (prices)
1970/71	950	5.3	14.4	3.2	8.6	4.5	12.1
1973/74	1170	10.4	17.3	5.5	9.1	7.9	13.1
1976/77	950	29.2	35.9	12.0	14.8	17.5	21.6
1979/80	1110	19.4	17.0	12.5	10.9	18.0	15.7
1981/82	1250	53.3	23.6	24.3	10.7	35.3	15.5

1. Partly extrapolated on the basis of the 1981/82 figures.
Sources: Oficina del Café (1983c), Aguilar et al. (1982)

Table 5.10. Coffee factories and processing and storage capacities (in 1000 t), by province (1981).

Province	Number of factories	Processing capacity (1982)	Storage capacity (1981)	Total quantities processed	
				1980/81	1981/82
San José	22	31	20	31.9	28.9
Alajuela	31	41	26	43.5	36.1
Cartago	24	24	21	19.0	15.7
Heredia	17	28	33	17.8	13.5
Guanacaste	5	1	1	1.7	1.8
Puntarenas	7	7	5	7.5	7.8
Limon	–	–	–	0.3	0.3
Total	106	132	106	121.7	104.1

Sources: Oficina del Café (1983c), Oficina del Café – MAG (1983)

processing factories, which are called 'beneficios'. The relations and transactions between producers, factories (beneficiadores) and exporters are all regulated by law. Producers have to deliver their coffee (cherries) within 24 hours after harvesting at the factory through its collecting agents or stations. They receive only part of the payment at the time of delivery, the remainder following several months later when final prices and margins in coffee trade are determined on the basis of the export prices and the information on costs submitted by factories and exporters. Producers may, however, receive advances from the factories. The 'beneficios' are to deliver part of their production, in recent years about 13-14 %, to the Oficina del Café. That amount is locally roasted and reserved for the domestic market. In Table 5.10 the provincial distribution of processing factories and their capacities is presented.

A typical coffee factory employs about 10 permanent staff and some 40 seasonal labourers in the three month period following harvesting. In some areas the processing starts as early as October, but in most areas it is concentrated in the period January to March. About 25 % of the processing factories are cooperatively owned and those had about 40 % of total processing capacity in 1982.

5.5.2 Processing costs

Every year a survey is carried out on the costs incurred by processing factories. Those studies are carried out on the basis of a sample, stratified according to size of factory and type of coffee processed; small: less than 23 000 double hectoliters (or about 530 t green coffee); medium: between 23 000-50 000 Dhl; and large: over 50 000 Dhl. The seven types of coffee considered are: Atlantics (3 subtypes), Hard Bean, Medium Hard Bean, Good Hard Bean and Strictly Hard Bean.

The studies on the processing of the 1980/81 harvest indicate that the variation of total costs is greater between the respective types of coffee than between different sizes of factories. With regard to size of factories it is, however, important to mention that with total costs for all three sizes around 4000 colones (US$220) per tonne, the costs of labour were higher for the small plants and the costs of depreciation, interest and transport were higher for medium and large plants. This could either be related to quality aspects, or it could be a matter of over-investment by those medium and large factories.

In Table 5.11 the costs of processing by type of coffee are reviewed. The processing costs are lowest for Hard Bean and Good Hard Bean Coffee and highest for Atlantics. The high processing costs of Atlantics coffee is mainly due to the relatively high costs of administration and transport. This is due to the relatively more dispersed cultivation of coffee in the Atlantic areas, with larger distances to factories and/or a less developed infrastructure.

For the determination of the final price to be paid to producers by the respective 'beneficiadores' the latter are allowed by law to deduct certain costs (including some material inputs, fuel, oil, electricity, insurances and social security and transport) from their total costs, up toa maximum of 15 colones per 46 kg green coffee (330 colones per tonne). Accounts of them have to be submitted. Without evidence the maximum deduction is 8 colones per 46 kg green coffee. Another law establishes the operational margin for the processing factories, fixed at 9 % of the price (f.o.r.) obtained by the factories, minus the processing taxes and the above mentioned 'special' costs. With an average price (f.o.r.) obtained by processing factories in 1980/81 of about 41 200 colones per tonne green coffee, and deductions of 5 and 330 colones per tonne respectively, the margin allowed for the factories was 3 680 colones per tonne. Although some price variations exist between the respective coffee types, the factories in the 'Atlantics'-region and to a far less extent those processing the Strictly Hard Bean type of coffee, were undoubtely in a difficult financial position in that year.

Table 5.11. Costs of coffee processing by type of coffee (colones/t) (1980/81).

Cost elements	Type of coffee[1]					Average of total (95)
	Atl. (19)	HB (17)	MHB (10)	GHB (11)	SHB (38)	
Processing						
wages, labour	750	780	830	630	720	740
material input	200	180	220	220	210	210
fuel, oil, electricity	570	460	380	480	450	450
subtotal	1520	1420	1430	1330	1380	1400
Factory						
insurance and social security	290	360	360	260	300	310
maintenance	390	330	540	250	260	350
depreciation	380	140	260	220	230	240
transport	650	220	440	350	270	370
subtotal	1710	1050	1600	1080	1060	1270
Administrative						
administration	600	360	190	260	320	300
taxes	190	40	110	110	380	170
office costs	150	60	90	60	50	80
subtotal	940	460	390	430	750	550
Finances						
interest on invested capital	530	250	440	460	770	500
interest on loans	360	160	160	320	240	240
subtotal	890	410	600	780	1010	740
Total costs	5060	3340	4020	3620	4200	3960 (US$ 220)

1. Abbreviations are given in Table 5.2. Number of coffee factories between brackets.
Source: Oficina del Café (1983b)

5.6 Marketing and pricing

5.6.1 Marketing channels and marketing functions

Because of the central position of processing factories, supported by law through detailed regulations, the marketing of coffee in Costa Rica is straight forward. The producers sell their cherry coffee to the cooperative or private factories usually through collecting agents (recibidores), and the factories, after having remitted a certain amount of coffee to the Oficina del Café for the domestic market, sell the coffee either directly or through exporters to foreign markets. The quantity reserved for the domestic consumption is sold through auctions, either directly to local roasting firms or indirectly through traders (Rodriguez Muñoz, 1980). The roasted coffee is sold through wholesalers and/or retailers to consumers. In recent years some roasted coffee has been exported. Apart from fixing the quota destinated for the local market, the Board of Directors of the Oficina del Café also fixes the proportion of exports to be sold to countries that are not member of the International Coffee Organization, in accordance with export-quota regulations established by that organization.

5.6.2 *Marketing margins and taxation*

The detailed marketing regulations and the indirect control over processing factories make it relatively easy to determine marketing margins and to administer and collect the respective production, processing and export taxes.

The major taxes on coffee in Costa Rica are the ad valorem production tax, and the ad valorem export duty. The first tax is to be paid by producers but is collected through the factories and amounts to 10 % of the free-on-rail price minus the processing tax and minus certain costs of processing materials to be declared by the factories (special costs). The ad valorem export duty ranges by law between 5 and 18 %, and constitutes a major element for domestic price stabilization. The major export tax was increased in 1980 and now amounts to at least 0.75 % of the f.o.b. value. There are also some minor export taxes and a tax on domestic coffee consumption, which constitutes the major source of income of the Oficina del Café. These taxes amount to about US$10 per tonne and about 43 colones per tonne green coffee, respectively. On top of these taxes a levy is charged of US$0.26 per 60-kg bag of coffee, which is paid to the International Coffee Organization (Certificate of origin-levy). Finally there is also a small sales tax of about 4.35 colones per tonne of coffee, (green coffee equivalent) delivered to the processing factories.

In Table 5.12 marketing margins and taxes are presented per tonne of green coffee. Since the average producer price depends on export as well as domestic prices, the price structure is shown for both the export market (above) and the domestic market (below). Over the period 1977/78 to 1981/82 the average unit sales value (export and domestic market) per tonne of green coffee amounted to US$2970 which was distributed as follows: producers: 69 %; Government through tax revenues: 17 % (of which 1.5 % to Oficina del Café); beneficios: 9.5 %; roasters: 1.5 %; exporters: 2 %, and wholesalers/retailers: 1 %.

5.6.3 *Export and domestic consumption*

In recent years about 13 % of the total coffee production was officially retained for the domestic consumption (Table 5.13). This amount of coffee is equivalent to about 6 kg green beans per person per year. Taking into account this high level of per person consumption and the low domestic prices of coffee, compared to prices on the world market (Table 5.12), the amounts reserved for the local market seem to be high. This rate of retention and the present price differential between

Table 5.12. Prices, marketing costs and margins in US$ per t (green coffee equivalents) (1977-1982).

	1977/78	1978/79	1979/80	1980/81	1981/82*
Exchange rate (colones per US$)	8.57	8.57	8.57	18.00	33.00
Percentage of coffee exported (%)	86	87	87	87	86
Export unit value (f.o.b.)	3900	3010	3910	2560	2640
Costs and margins					
export taxes	10	10	10	20	30
ad valorem duties	300	308	350	200	285
certif. of origin levy	5	5	5	5	5
costs of exporting	40	40	40	40	40
other	30	1	140	4	1
Total	385	364	545	269	361
Unit value obtained by factories (f.o.r.)	3515	2646	3365	2291	2281
Costs and margins					
processing taxes	1	1	1	1	1
costs allowed for processing	20	20	20	20	20
production taxes (10 % of f.o.r.)	350	260	340	230	230
margin for processing	310	240	300	210	210
Total	681	521	661	461	461
Unit value to producers for export coffee	2834	2125	2704	1830	1820
Producer price	2510	1940	2450	1670	1640
as share of export unit value (%)	64	64	63	65	62
Unit value to producers for domestic coffee	570	690	780	415	395
Costs and margins					
margins for beneficios	250	270	330	240	210
consumption taxes	5	5	5	5	5
margin for roasters	220	310	370	310	280
margin wholesale & retail	115	145	205	190	170
Total	590	730	910	745	665
Domestic consumer price	1160	1420	1690	1160	1060

Source: Oficina del Café (1981, 1982, 1983c)

export and domestic prices can, however, constitute useful instruments for the stabilization of producer prices. The coffee sold at the domestic market should contain at least 20 % high-quality coffee.

Coffee exports increased gradually over the period 1970 1983, and because of low stocks, Costa Rica was not able to increase exports during the price boom in 1976/77 (Table 5.13). In the last few years stocks increased again and probably constituted 50 % of annual production in 1983/84. Coffee is exported all year around without major fluctuations. The major export harbours are Limon, Penas blancas and Puntarenas, in 1980/81 handling 70 %, 21 and 7 % of total exports, respectively. The major destinations of Costa Rica coffee are the United States,

Table 5.13. National production, domestic consumption and exports (in 1000 t) (1970-1982).

Crop year	Opening stock	Production	Consumption	Export	Ending stock
1970/71	12.4	77.7	6.6	61.2	20.2
1971/72	20.2	93.1	6.3	73.0	31.4
1972/73	31.4	76.0	7.9	87.4	10.8
1973/74	10.8	92.3	9.6	80.2	13.4
1974/75	13.4	85.8	9.3	84.7	5.3
1975/76	5.3	76.6	10.4	61.0	10.6
1976/77	10.6	79.9	10.9	72.9	6.5
1977/78	6.5	86.9	11.1	76.3	6.3
1978/79	6.3	104.9	13.0	91.9	6.5
1979/80	6.5	91.3	12.7	79.5	5.8
1980/81	5.8	128.4	12.5	93.9	27.9
1981/82	27.9	96.5	14.6	93.0	16.9
1982/83	16.9	138.0	13.4	102.0	39.5

Source: USDA (1983)

Table 5.14. Proportional distribution (%) of value of coffee exports to importing countries (1977/78-1981/82).

	1977/78	1978/79	1979/80	1980/81	1981/82
ICO members	92	95	98	87	88
United States	20	29	28	14	18
West Germany	30	18	20	25	21
Netherlands	12	11	12	11	9
Finland	8	9	11	9	8
others	22	28	27	28	32
Non-members	8	5	2	13	12
East Germany	0	–	0	5	4
Czechoslavakia	1	1	–	3	2
Romania	3	2	–	1	0
others	4	2	2	4	6

Source: Oficina del Café (1983c)

West Germany, the Netherlands and Finland among the ICO-member-countries, and East Germany, Czechoslovakia and Romania among the non-members (Table 5.14). Sizeable amounts of semiroasted coffee were in 1981/82 sold to Puerto Rico.

5.7 Linkages and supporting services

5.7.1 Research and extension services

Research related to coffee production and processing is carried out by cooperative programmes between the Oficina del Café and the Ministry of Agriculture, and between the Oficina and the Universidad de Costa Rica. The Centro de Investigaciones en Café (Cicafé) is particularly involved in the selection of plant material and in coffee technology and the diffusion thereof.

Under the joint programme with the Ministry of Agriculture studies have been carried out through the following projects (with results obtained in 1980/81).

– Mineral fertilization: the application of 300 kg pure nitrogen per hectare (half as nitrogen and half as compound (N-P-K-Mg-B: 18-5-15-6-2) fertilizer) in three applications appeared the most promising, under otherwise optimum conditions.

– Genetic improvement: the variety Catuai showed the highest yields followed by Caturra and Mundo Novo; in collaboration with IICA (Promecafé) and CATIE varieties and hybrids were identified which show resistance to leaf-rust. (Not yet distributed in 1983 because of fear that farmers might stop treatments, which might cause return of other diseases.)

– Cultivation practices: research has been carried out on optimum planting distances, and pruning systems.

– The application of herbicides: the type of herbicides, the quantities to be applied during the respective production stages and the methods of application were studied. The herbicides: Lazo (480 g/l) and Goal (240 g/l) proved the most effective in liquid form at a rate of 2.8 l/ha.

The activities undertaken under the joint programme with the University include several diversification projects in the coffee zones and studies on agro-ecological suitability and production costs of various crops. Attention has been given, among others, to: strawberries, cardamon and grapes. A diversification programme is also undertaken resulting from the International Coffee Agreement, which has paid attention among others to the introduction of macademia nuts. The total costs of the two cooperative programmes Oficafé-MAG and Oficafé-UCR amounted in 1980 to respectively 3.2 million and 0.4 million colones.

The agricultural extension services were established in 1948 after the reorganization of the Ministry of Agriculture and with the collaboration of STICA. Extension activities for coffee production were further enhanced after the establishment in 1962 of the Ministry of Agriculture-Oficafécooperative programme. Major activities are demonstration plots (about 30 every year), theoretical and practical courses (about 10 per year with an attendance of about 40 farmers), practical training in the field for groups of farmers and personal visits to farmers. Through these activities about 3500 farmers are reached yearly or about 10 % of all coffee farmers.

5.7.2 Provision of credit and payment regulations

The payment to farmers for coffee delivered to the processing factory takes place in two installments. The first of these or provisional payment should be paid ultimately before the 31st of March after the harvest and depends on the quantities already sold for export. The final producer price is determined by the Junta de Liquidaciones of the Oficina del Café for every individual processing factory in the country in the month of October, on the basis of the documents provided by the factories and by the exporters.

Because of these delayed payments farmers do need credit facilities, which are usually provided by the 'beneficios' or directly by banking agencies and the 'Juntas Rurales de Crédito Agricola'. In the latter case farmers are obliged to inform these agencies of the debts they owe to the factory. The agencies are not allowed to charge farmers a higher interest rate than coffee factories, whereas the latter may charge farmers an interest rate 1 % above the rate they have to pay themselves. The farmers are not considered to pay the interest on prepayments (before delivery) received from the factory, as from the date the calculation of provisional payments has been made. Although described in details, these regulations are not always abided to, in practice.

Of the total amount of agricultural credit supplied by the National Banking System of 3045 million colones in 1980, about 20 % was for coffee production. Interest rates on credit for coffee, sugar-cane and livestock activities were 12 %, against 10 % for most other agricultural activities.

5.7.3 Input supply

The major farm inputs in coffee production are seedlings, fertilizers, insecticides and herbicides. To accelerate the technological change from traditional to modern cultivation practices, the Oficina del Café has been actively involved in provision of selected seed mainly consisting of Caturra and to a lesser extent Catuai varieties. The organization also superfices every year more than 200 nurseries, which on average produce about 45 000 seedlings.

The intensification of coffee production systems brought about a steep increase in the use of chemical fertilizers. Whereas before 1950 farmers mainly applied manure and organic material and only little fertilizers, in 1981/82 farmers used on average not less than 700 kg fertilizers per hectare (most of which compound fertilizers with 18 % nitrogen). This accounts for about a third of the national consumption of fertilizers which amounted in 1981/82 to about 45 000 t N, 12 000 t P_2O_5 and 20 000 t K_2O. A large part of fertilizers, fungicides, insecticides and herbicides is supplied to farmers through the cooperatives and the processing factories, usually on credit.

5.7.4 *Use of byproducts*

Coffee pulp has traditionally been used as fertilizer on coffee farms, but has gradually been replaced by commercial fertilizers and now presents in some areas a pollution problem. The concentration of processing activities in central factories clearly discourages the use of pulp on the farm, but it offers opportunities for industrial manufacturing.

Fresh and ensiled pulp can be used as an ingredient of animal feed. Fresh pulp is to be processed within 24 hours after pulping, and is fermented after concentration and the application of lime. Subsequently it is pressed, dehydrated, pulverized and packaged. Ensiled pulp is first pressed and then placed in suitable silos where the extracted juice is added again, with a 2 % molasses solution and some forage. The process is allowed at least 100 days of anaerobic fermentation. After being ensiled and following a bacteriologicl examination, it is treated in a similar manner as fresh pulp.

A plant in the central valley of Costa Rica with a capacity of 40 000 tonne of pulp (more than 20 % of national supply) required a total investment of US$2 million (1979) and provides permanent employment for 25 people. The company studies the possibility of using coffee hulls as fuel for its burners, which could reduce fuel cost by nearly 50 %.

5.8 Summary and economic parameters

Costa Rica is one of the Central American countries, whose national economy for a large extent depends on the production of coffee and bananas. In its Central Highlands, where most of the population resides, the soils of volcanic origin and the temperate climate with about 9 humid months offer excellent conditions for cultivation of coffee.

In the past decade the coffee production has expanded into new areas, at lower altitudes, where generally lower-quality coffee is produced. But at the same time a further intensification of coffee cultivation took place through the further introduction of the compact varieties Caturra and Catuai. The national average yield of 1300 kg per hectare is among the highest in the world. Now nearly all coffee produced in the country is of the arabica type, and in the trade is classified under 'other milds'. Because of the recent incidence of leaf-rust, further emphasis is given in research to varieties that show resistance to that disease.

All processing activities are undertaken in about 100 factories, which also perform a central role in the marketing of coffee and in provision of credit to producers. The national coffee institute, 'Oficina del Café' exercises extensive control over the coffee sector, through detailed regulations, pricing and taxation. To fix prices and margins it every year carries out studies on the costs of production and the costs of processing and marketing.

The government, well aware of the danger of relying on a few main exports, has

Economic parameters of the coffee sector in Costa Rica (1982).

Resources used in coffee sector	
Land under (productive) coffee	85 000 ha
as share of arable & permanent crop land	17 %
Labour involved[1]	60 000 man-years
in production	50 000 man-years
in processing, marketing & services	10 000 man-years
as share of national work force	5 %
Material inputs for coffee production	1 000 million colones*
Annual import requirements for coffee industry	25US$ million*
Income from coffee sector	
Value of coffee production	
(volume × export unit value)	9 000 million colones
as share of agricultural production	20 %
as share of GDP	5 %
Value of coffee exports (f.o.b.)	238US$ million
as share of agricultural exports	36 %
as share of exports	27 %
Share of total production value, obtained by	
Producers	69 %
Processing industries, private traders & exporters	14 %
Government and parastatal organizations	17 %
Other parameters	
Arabica (other mild) production	100 %
Yield (green coffee)	1 300 kg/ha
Production of labour involved (green coffee)	5.8 kg/man-day
Share of coffee produced for export	86 %
Ratio producer price (per kg green coffee):	
daily wage rate (colones/man-day)	0.75

1. man-year = 300 man-days.
Sources: FAO (1971-1983), IMF (1983), World Bank (1984), and own estimates

initiated several diversification projects, but these have not yet shown satisfactory results. And it seems difficult to stick to the policies of not increasing the area under coffee and not growing coffee in less suitable zones (and thereby affecting the good name of Costa Rica coffee).

References

Aguilar, F.J., Barboza, V.C. & Leon, S.J., 1982. El desarollo technologico del café en Costa Rica y las politicas científico technologicas. CONICIT, San José.

Banco Central de Costa Rica, 1980. Cifras sobre producción agropecuaria, 1974-1979. San José.

Banco Central de Costa Rica, 1983. Informe sobre la política de crédito a la actividad cafetalera durante 1982. San José.

Campos, C.F., 1982. El cultivo del café. San José.
Dirección General de Estadística y Censos, 1974. Censo Agropecuario, 1973. San José.
Dijeres Morales, C.E., 1981. La compra-venta del café para el consumo nacional. Universidad de Costa Rica (Tesis), San José.
FAO, 1971-1983. Production Yearbook. FAO, Rome.
Hall, C., 1982. El café y el desarrollo histórico-geográfico de Costa Rica. San José.
IICA-Promecafé, 1982. Bibliografía del café. Turrialba.
IICA-PROTAAL, 1982. Políticas publicas y cambio tecnico entre les pequenos productores de Pejibaye, Perez Zeledon, Costa Rica: Una interpretación tentativa con base en información parcial. San José.
IMF, 1983. International Financial Statistics. IMF, Washington.
KIT, 1974. Landendocumentatie Costa Rica. KIT, Amsterdam.
Oficina del café, 1976. Tipos y subtipos de café en Costa Rica. Oficina del café, San José.
Oficina del café, 1977. Análisis económico de la producción de café en Costa Rica; Cosecha 1976-1977. Boletin tecnico no. 8. San José.
Oficina del café, 1981. Leyes y reglamentos usuales sobre café. San José.
Oficina del café, 1982. Informe de labores, 1981. San José.
Oficina del café, 1983a. Análisis económico de la producción de café en Costa Rica. Cosecha 1981-1982. Boletin technico no. 25. San José.
Oficina del café, 1983b. Investigación económica de los costos de beneficiado en tres tamanos de beneficio y siete tipos de cafe. Cosecha 1980-1981. Boletin technico no. 26. San JoseB.
Oficina del café, 1983c. Informe sobre la actividad cafetalero de Costa Rica. San José.
Oficina del café-MAG, 1982. Manual de recomendaciones para cultivar café. San José.
Oficina del café-MAG (Programa cooperativa), 1983. Informe annual de labores, 1982. San José.
Rodriguez Munoz, G., 1980. Universidad de Costa Rica (Tesis). Análisis parcial del subsector cafetalero con énfasis en crédito y comercialización. San José.
USDA, 1983. Forteign Agriculture Circular on Coffee. USDA, Washington.
World Bank, 1984. World development report. Oxford University Press, Washington.

Currency equivalents and price indices in Costa Rica (1960-1983).

	Exchange rate (colones per US$)	Consumer price index (1980 = 100)
1960	5.6	28
1965	6.6	32
1970	6.6	36
1975	8.6	68
1976	8.6	70
1977	8.6	73
1978	8.6	78
1979	8.6	85
1980	8.6	100
1981	21.8	137
1982	37.6	261
1983	41.1	346

Source: IMF (1983)

6 Coffee in Kenya

6.1 Background

Kenya is situated along the East African Coast and stretches from 4° South to 4° North of the Equator. It is surrounded by five countries: in the south and east, by Tanzania and Uganda, with which countries it formed the East African Community; untill 1977 and in the north by Sudan, Ethiopia and Somalia. The total land area is 580 000 km^2, of which about two-thirds receives little rain and is only sparsely populated. In the other areas, the population density is high, and particularly in Central, Nyanza and Western provinces with a population density of about 200 people per square kilometre. In mid-1982, the total population was 18.1 million, and over the past decade the average annual growth rate of the population was not less than 4 %. The capital Nairobi has nearly a million inhabitants.

Administratively the country is divided into seven provinces, and the city of Nairobi, and the provinces are further subdivided into districts, divisions and locations.

The GNP per person was in 1982 US$390. The share of the major sectors in the GDP in 1982 was agriculture 33 %, industry 22 % and services 45 %. And the shares in the labour force was in 1982 as follows: agriculture 78 %, industry 10 % and services 12 % (World Bank, 1984).

Table 6.1 shows the major crops in Kenya. Agricultural production is largely concentrated in the higher altitude zones (above 1500 m), which coincide with the most populated areas. About 58 % of total exports consists of agricultural and agro-industrial products, of which coffee and tea are the major commodities with respectively 22 % and 12 % of total exports (1980-1982, IMF-Statistics, 1983). With an average annual production of 84 000 t in the period 1979 to 1981, Kenya ranked 14th as world producer of coffee, but the quality of its coffee (Colombian mild arabica) is one of the highest in the world.

6.2 Ecology

Kenya is characterized by considerable contrasts in topography, climate and soils, and less than 30 % of the total area has agricultural potential (excluding rangeland). The altitude ranges from sea level to over 5000 m (Mount Kenya), and increases gradually from below 200 m in the east to around 2000 m in the central and south western areas (Figure 6.1). Around Lake Victoria the altitude is

Table 6.1. Average area harvested, yield and national production of major crops over 3-year periods.

Crops	Area (1000 ha)		Yield (kg/ha)		Production (1000 t)	
	1969/1971	1979/1981	1969/1971	1979/1981	1969/1971	1979/1981
Food crops						
maize	1383	1240	1489	1580	2060	1939
wheat	133	119	1678	1779	223	212
sorghum	201	210	1070	994	215	209
beans & peas (dry)	565	550	472	426	267	235
Oil crops						
coconuts	.	.	.	.	77	90
sesame	8	20	401	406	3	8
sunflower	4	14	718	1095	3	15
Beverages						
coffee (green)	85	122	668	694	57	84
tea (made)	30	61	1246	1532	38	93
Other crops						
cotton	74	124	215	293	16	36
sugar-cane[1]	26	38	5692	11316	148	430
sisal	52	44	888	1025	46	44

1. Production of sugar centrifugal and non-centrifugal.
Source: FAO (1971-1983)

again somewhat lower: between 1000 and 1600 m. Arabica coffee is grown in Kenya at altitudes between 1400 and 1900 m. The differences in altitude have a pronounced influence on the average annual temperatures, which are above 25° in the eastern and northern parts of the country and below 20° in the hilly and mountainous areas of Rift Valley and Central Provinces. The average annual rainfall follows a similar pattern with less than 500 mm in the north, in the east and in some areas in the south and between 500 to 1500 mm in the central and south-western areas. There is also a small strip along the coast, with more than 500 mm and in some places more than 1000 mm (Figure 6.2). An average annual rainfall of more than 1500 mm is only found along the mountain ranges and in Western Province. The dry period is in several areas from December to February followed by the long rains (March to June) and often by a short rainy period in September to October (Table 6.2). The minimum amount of rainfall required for coffee growing in Kenya varies from 1000 to 1150 mm depending on whether there are two rainy seasons or only one. In some areas coffee is grown under conditions of minimum rainfall; then mulching is essential to conserve moisture, or irrigation is required. In the Ministry of Agriculture's Farm Management Handbook of Kenya (1983) agro-ecological zones are definied on the basis of temperatures (related to altitude) and rainfall characteristics (precipitations and evaporation). The re-

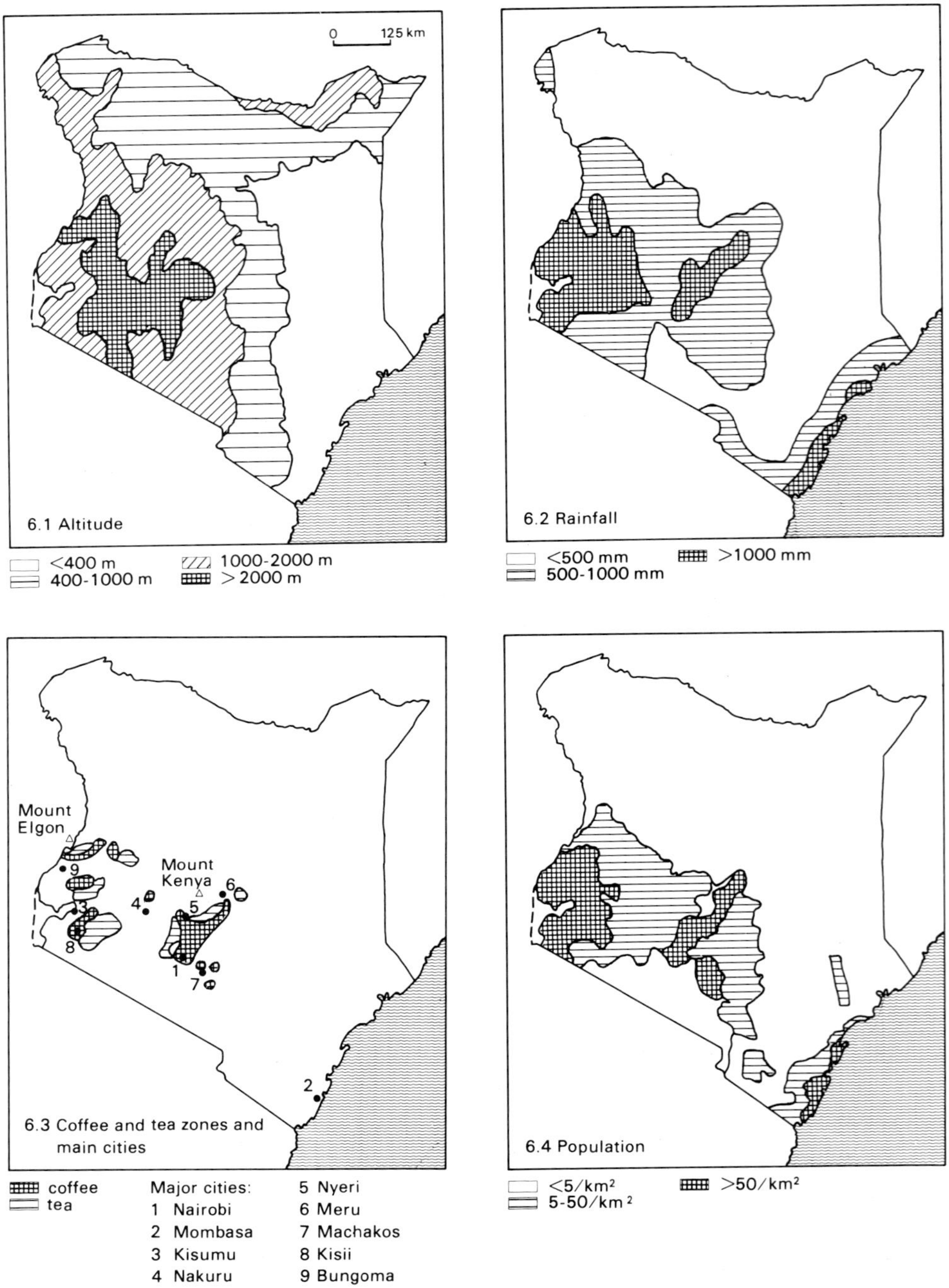

Fig. 6.1-6.4. Altitude (6.1), average annual rainfall (6.2), coffee and tea zones and major cities (6.3) and population density (6.4). After Jomo Kenyatta Foundation (1982).

Table 6.2. Distribution of rainfall for several stations.

Station	Altitude (m)	Annual rainfall (mm)	Monthly rainfall (mm)					
			J	F	M	A	M	J
Central Province								
Kiambu	1767	1014	46	46	98	226	168	53
Nyeri	1829	905	51	46	68	167	157	30
Rift Valley Province								
Nakuru	1836	844	16	24	60	127	107	80
Kitale	1890	1182	21	43	70	143	159	133
Nyanza Province								
Kisii	1740	1785	70	89	176	265	220	146
Western Province								
Kakamega	1554	1929	60	95	151	254	260	190
Eastern Province								
Machakos	1640	899	52	51	124	205	74	12
Meru	1555	1602	58	24	106	353	150	10
			J	A	S	O	N	D
Central Province								
Kiambu			28	25	35	67	139	83
Nyeri			37	38	33	92	109	77
Rift Valley Province								
Nakuru			109	110	67	54	60	30
Kitale			155	175	105	81	62	35
Nyanza Province								
Kisii			111	151	171	138	141	107
Western Province								
Kakamega			164	230	186	134	118	87
Eastern Province								
Machakos			5	6	8	53	189	120
Meru			18	19	11	317	375	161

Source: Jaetzold & Schmidt (1983)

spective zones have been named after the potentially leading crops. Depending on rainfall three different ecological zones for coffee are distinguished: the coffee-tea zone (UM1), the main coffee zone (UM2) and the marginal coffee zone (UM3).

Because of the limited water supply and relatively long dry seasons in most areas the soils require a considerable depth (2 to 3 m) for water storage capacity. The best soils for coffee in Kenya are those derived from recent volcanic deposits. The predominant type in Central Province is the rich, well drained, deep, dark reddish brown, friable clay of the Kikuyu area, known as 'Kikuyu red loam' (Nitosols). Similar soils, based on igneous rocks (basalts) and volcanic ash are found around Mount Kenya (humic Andosols) and in Kisii district (Nitosols and Phaeozems). In other coffee areas, in Rift Valley and Western Provinces, coffee is generally grown on less fertile soil types (Ferralsols and Acrisols).

Taking into account the agro-ecological and soil conditions in the country and the actual areas under coffee, it appears that there is not much scope to expand the area under coffee. Coffee is in fact already grown in less suitable areas and on less fertile soils and because of the high population density in many coffee areas, much emphasis is also laid on food crop production. Some attention is now being given to the increse of robusta coffee plantings in the areas around Lake Victoria (Siaya and Busia Districts).

6.3 The coffee sector

6.3.1 *History*

Coffee was introduced into Kenya by French missionaries in about 1894. The coffee was first planted at their mission station in Bura, from where seed was taken to the St. Austin's Mission in Nairobi. The introduction of coffee by another mission at Kibwezi was less successful. At the St. Austin's Mission nurseries were established from where seedlings were supplied to large farms in the Kikuyu area and by 1908 the first commercial plantations were established (Waters, 1969).

The industry initially developed along estate lines and by 1935 more than 40 000 ha were planted with arabica coffee, producing about 22 000 t of washed coffee. Due to poor prices, the area (ha) declined in the period 1935 to 1950 steadily to 24 000 ha, producing only about 10 000 t, but the sharp rise in prices from 1950 onwards brought production back to the level of 22 000 t in 1957. Except for some trials in smallholder areas of Meru and Kisii-districts coffee was only grown by expatriate farmers, but following the Emergency and by the Swynnerton plan of 1946, smallholder growing of coffee was initiated in 1952, in all suitable areas in the country. By 1966 the production by smallholders overtook that by the estates. The further expansion of the smallholder coffee production was then hampered by two important factors: the restriction on production and sales imposed by the International Coffee Agreement in the period 1964 to 1972 (and a ban on new plantings in Kenya) and the incidence of coffee berry disease which affected smallholdings more than estates. Smallholder production increased again from 1973 onwards and since 1977 remained ahead of the estate production.

6.3.2 *Recent developments*

Since 1978 the smallholder sector contributes about 60 % of the national coffee production. The area (ha) of smallholder plantations is even three times as high as that of estates, which reflects the disparity between the average yields obtained by the two sectors. Table 6.3 shows the national coffee production, by the two sectors, over the period 1950 to 1983. The Table illustrates that in comparison with the fairly constant production of the estate sector, the development of the

Table 6.3. National production of coffee (1000 t) by smallholder cooperative and estates (1950-1982).

Crop year	Smallholder coop's washed coffee	Smallholder coop's hulled mbuni	Estates washed coffee	Estates hulled mbuni	Others	Total production	Proportion by estates (%)
1950/51	0.1	–	9.8	–	0.1	10.0	98
1960/61	7.9	–	25.4	–	0.3	33.6	76
1965/66	25.5	–	25.7	–	0.9	52.1	49
1970/71	26.3	–	28.6	–	5.0	59.9	48
1975/76	32.4	3.7	36.3	1.4	0.8	74.6	51
1976/77	44.5	3.2	48.4	1.3	3.8	101.2	49
1977/78	43.9	3.9	32.1	1.6	3.5	85.0	38
1978/79	41.8	4.3	26.0	0.8	1.4	74.3	36
1979/80	46.7	5.2	37.3	1.8	0.7	91.7	43
1980/81	54.5	9.5	33.0	1.7	1.0	99.7	35
1981/82	43.9	8.6	32.6	1.8	0.5	87.4	39
1982/83	44.5	8.0	30.5	2.5	0.6	86.1	38

Source: Coffee Board of Kenya (1982)

Table 6.4. Area planted, national production and average yield of coffee by Province (1981/82).

Province	Smallholders area (1000 ha)	Smallholders production (1000 t)	Estates area (1000 ha)	Estates production (1000 t)	Average yield (kg/ha)
Eastern	45.0	18.9	2.2	1.8	420
Central	38.6	27.9	24.1	31.0	922
Rift Valley	0.4	0.1	7.3	1.6	203
Nyanza	8.6	4.0	–	–	482
Western	4.5	1.6	–	–	378
Coast	0.4	0.1	–	–	250
Total	97.5	52.5	33.6	34.4	
Average yield (kg/ha)					
1980/81	756		1057		848
1981/82	539		1023		667

Source: Compilation from Coffee Board of Kenya (1982)

smallholder sector took place in three stages: a fast increase of production from 1955 until 1965, a period of quasi stagnation from 1966 to 1976, and since 1979 a rapid increase of the area planted, which due to declining yields, only resulted in a minor increase in production. The proportion of mbuni (unwashed) coffee increased in this last period.

The regional distribution of coffee production also varies by farmtype (Table

6.4). The coffee estates are heavily concentrated in Central Province, and particularly in and around Kiambu District. Eastern Province (Meru District) has on the other hand the largest area of smallholder coffee, but because of the higher yields, Central Province has also the largest smallholder production. On the western side of the Rift Valley important smallholder coffee areas are found in Kisii District (Nyanza Province) and in some areas of Western Province, whereas some coffee estates are found in Rift Valley Province.

Whereas average yields for estates are above 1000 kg/ha, smallholder yields reach on the average only about 700 kg/ha. This is partly due to the occurrence of neglected holdings, and partly to a lower level of inputs (fertilizers, chemicals, etc.). To upgrade the smallholder coffee production, the World Bank-funded Smallholder Coffee Improvement Project (SCIP) is in operation since 1980. The Coffee Research Station has developed some new disease resistant, compact, varieties, which will be issued in 1986, and may double average yields per hectare.

6.3.3 Organization

Coffee is in Kenya produced by about 700 estates and some 320 000 smallholders, which are grouped in about 150 cooperative societies. The pulping, fermenting and drying of the coffee is carried out by the estates and the cooperatives themselves. After this processing they deliver the parchment coffee to the KPCU (Kenya Planters Cooperative Union). This organization takes care of the hulling, on behalf of the Coffee Board of Kenya, and after grading and bagging, the coffee is sold to about 50 exporters through public auctions, held weekly.

Under the Coffee Act of 1933, the Coffee Board was established to supervise and regulate all sections of the coffee industry and to represent the Government in matters pertaining to international agreements in relation to coffee. The main activities of the Board relate to: the licensing of producers, pulping factories, exporters and other participants in the coffee trade; the classification of all coffee; the marketing and payment to growers; the insurance of the total coffee crop and all aspects of publicity.

Research on coffee is carried out by the Coffee Research Foundation, which is largely financed by the Coffee Board through funds accruing from the statutory levy on sales of coffee. The coffee extension services are the responsibility of the Ministry of Agriculture, but these are supplemented by advisory officers from the Research Station and the Board's Inspectorate Services. The Ministry of Cooperative Development supervises and assists the primary societies and cooperative unions in provision of inputs and loans to coffee growers, in factory management and administration, and in the education of cooperative members and personnel.

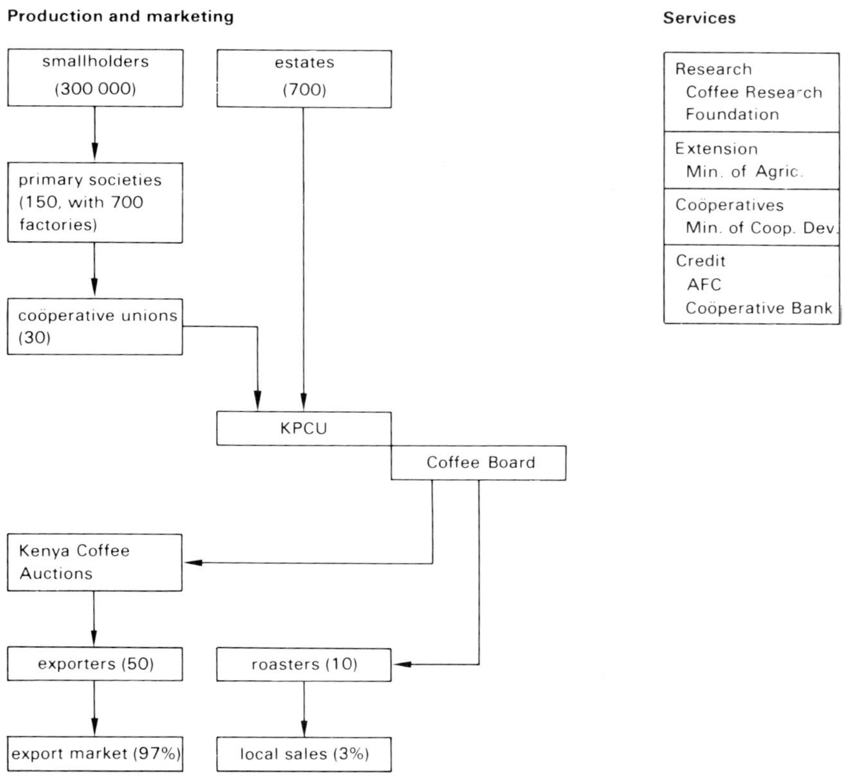

Fig. 6.5. Participants in the coffee sector and main marketing channels in Kenya in 1982.

6.4 Production

6.4.1 Production units and farming systems

Agriculture in Kenya has a pronounced dual character, with on the one hand the small peasant farmers forming the bulk of the farming population and on the other hand the large commercial farms. Both sectors cover a total area of roughly 3 million ha, but in the smallholdings this is shared by about 2.7 million farming families, whereas the number of estates is less than 4000 (Table 6.5). Little attention has generally been given to the third category of so-called 'intermediate farms', ranging from approximately 8 to 50 ha, with a total of 53 000 farms and a total area of about a million ha (1979).

Large farms are for historical reasons heavily concentrated in Rift Valley and Central Provinces, whereas the smallholdings are more evenly spread, with the highest number in Nyanza Province. The average farm size among smallholders is about 1 ha or less in Central, Nyanza, and Western Provinces and about 15 ha in the other provinces. The average size of estates is about 700 ha, with more than 50 % being less than 200 ha and about 6 % exceeding 2000 ha.

Major smallholder crops are: maize (staple food), pulses, sorghum, rootcrops and the cash crops: coffee, tea, cotton and pyrethrum (Table 6.6). Livestock, and particularly dairy products are also important. Estates are particularly involved in

Table 6.5. Number of farms by category and by provincie, and their proportional distribution by size (in %) (1979).

Farm size (ha)	Province							Total area (1000 ha)
	Western	Nyanza	Rift Valley	Central	Eastern	Coast	total	
Small holdings								
less than 2	63	74	66	69	60	58	67	
2-4	32	23	20	27	33	31	26	
more than 4	5	3	14	4	7	11	7	
number (× 1000)	342	708	527	501	456	160	2694	(3216)
Intermediate farms number (× 1000)								
8-50	4	5	29	4	10	1	53	(1040)
Large farms								
50-200	36	78	44	75	46	49	54	112
200-2000	45	18	51	22	23	35	40	895
more than 2000	19	4	5	3	31	16	6	1652
number	11	116	2327	1067	163	51	3735	(2569)
All farms								
number (× 1000)	346	714	558	506	466	161	2751	(6915)

Source: Central Bureau of Statistics (1981, 1982)

Table 6.6. Proportion (%) of smallholders growing selected crops by province (1974/75).

	Western	Nyanza	Rift Valley	Central	Eastern	Coast	Kenya
Local maize	74	80	59	95	99	94	86
Hybrid maize	73	36	92	67	30	19	50
Beans	79	39	22	98	86	28	69
English potatoes	–	1	8	86	52	2	32
Sorghum	37	75	1	1	16	2	30
Coffee	5	21	5	45	44	1	27
Tea	4	–	15	18	11	–	12
Pyrethrum	–	18	16	8	7	–	9
Cotton	20	17	5	–	2	5	9

Source: Central Bureau of Statistics (1977)

livestock enterprises and in the production of wheat, maize, sisal, sugar-cane, coffee and tea.

Cultivation practices of coffee in Kenya were always both labour as well as capital intensive. The ample availability of low-cost labour, the need for regular spraying against pests and coffee berry disease and the necessity, because of low rainfall, to apply mulching and supplementary irrigation, all contributed to a high level of inputs in the estate sector. Because of these same factors the introduction of coffee growing into smallholdings was prohibited for a long time, by reason that it would reduce the labour supply for the estates and spread diseases, due to insufficient spraying. The first three districts in which coffee growing by smallholders was allowed in 1933: Meru, Kisii and Embu, were relatively far removed from European coffee growing areas. Smallholders had to follow well-established cultivation practices and had to adhere to a marketing cooperative. Although the introduction of coffee in the small-farm sector had probably some influence on the labour supply and the wage rates in the estate sector and certainly caused more problems in controlling pests and diseases, it did not damage the good name of Kenya coffee.

A typical smallholder in the main coffee areas with a 1.5 ha farm, grows respectively about 0.5 ha maize; 0.2 ha beans, potatoes or vegetables; 0.1 ha bananas and 0.2 ha coffee and he uses 0.5 ha for pasture and fodder for his dairy cows (kept often on zero grazing) and for mulch for his coffee. Since they rely for their cash income mainly on the sale of coffee (and milk) and since many of them have become used to hybrid maize and grade cattle, they accept high standards of cultivation for coffee, as long as the inputs are available and the price ratio inputs/outputs is favourable.

The smallholder coffee areas can be classified according to the ecological zones, suitable for coffee growing and to their actual location (Table 6.7). Major differences in cultivation techniques and varieties used exist between the ecological zones, but the vicinity to the estates among other factors also brought about

Table 6.7. Some characteristics of the major smallholder coffee zones, including proportional distribution (%) of coffee areas over the ecological zones.

Coffee zone	Coffee area (%)				Main variety	Proportion shaded (%)	Inputs (KShs/ha)	Yield (kg/ha)
	UM1	UM2	UM3	total				
Centr. Prov. & Embu district	22	25	6	53	SL28/34	3	1500	810
Meru district	8	14	7	29	SL28/K7	50	200	442
Machakos district	–	–	9	9	SL28	2	400	434
Kisii district	9	–	–	9	Blue Mt.	32	300	465
Total	39	39	22	100		19	900	640

Source: Coffee Research Foundation (1984)

Table 6.8. Some characteristics of irrigated and non-irrigated coffee estates by size group.

Size group (ha)	All estates		Irrigated			Non-irrigated		
	avarage size (ha)	proportion irrigated (%)	number of applications		yield (kg/ha)	number of applications		yield (kg/ha)
			ferti-lizers	fungi-cides		ferti-lizers	fungi-cides	
8-20	14	31	4	9	1300	2	6	650
20-60	41	57	3	9	1150	2	6	600
60-140	98	70	4	9	1300	2	8	900
more than 140	277	73	5	10	1400	2	9	900
Total	58	64	4	9	1300	2	7	850

Source: Coffee Research Foundation (1984)

differences between districts with similar ecological conditions. In smallholder areas far away from estates, a large proportion of coffee plantations is shaded, which is accompanied by a relatively low level of fertilization, and modest yields.

Estates can be classified by size and on the basis of whether they use irrigation or not (Roe & Whitaker, 1984). A great number of estates is located in areas with a relatively low annual rainfall and a pronounced dry season in the period June to September. Therefore not less than 64 % of the estates apply supplementary irrigation. On irrigated coffee estates, fertilizers and fungicides are applied more frequently, which also contributes to higher yields (Table 6.8).

6.4.2 Production stages and production costs

Whereas the annual costs of coffee production vary considerably by ecological zone and type of coffee plantations, the costs of establishment do not differ much between areas and farm types. Hereunder the activities and input requirements for the establishment of a coffee plantation are reviewed.

In the coffee nurseries, the seed, obtained from the Coffee Research Station or its substations, is first germinated in sandbeds of 5 to 7 cm depth, before being transplanted into 'polybags'. The seedlings are usually shaded and the maintenance includes watering, twice per week, and weeding, at least once a month. In a nursery of 0.1 ha about 25 000 seedlings can be raised, for the filling of the bags requiring about 10 m^2 manure, 30 kg single superphosphate and 16 kg Aldrin (2.5 %) dust (Coffee Research Foundation, 1983).

Meanwhile the land preparation takes place, including clearing, weeding and on sloping land, the establishment of soil conservation measures. Then planting holes are dug at dimensions of 60 cm × 60 cm, which are filled with manure (one debe of 20 l), 0.1 kg double superphosphate, 0.1 kg limestone and 50 kg Aldrin

(2.5 %)-dust. Instead of cattle manure, rotted coffee pulp is sometimes used. The conventional spacing of coffee plants is 2.75 m × 2.75 m, giving a plant population of about 1300 trees per hectare. Under irrigation the planting density is sometimes increased to around 2000 trees per hectare (2.25 m × 2.25 m or 1.75 m × 3 m). The new compact varieties (Catimor) to be released in 1986/87 will probably be planted at a spacing of 1.25 m × 2 m (4000 trees per hectare). Transplanting to the field takes place when the seedlings are 30 to 40 cm high, which is usually about 12 months after germination.

In Table 6.9 the estimated establishment costs per hectare are given for the traditional plantations and future plantations with the improved varieties. Because of the higher planting density the labour inputs of digging and planting are higher for the improved plantations, and the seedlings will become more expensive. But the establishment period will be shorter, reducing the total costs of maintenance until the trees reach maturity. Because of the infrastructure of buildings, irrigation and other equipment, the total capital investment for estates is high and amounted in 1983 to Kshs 40 000-50 000 per hectare (Roe & Whitaker, 1984).

Table 6.9. Establishment costs per hectare for traditional (small holder) plantations (1300 trees per hectare and 3 years establishment period) and for plantations with improved hybrid cultivars (4000 trees per hectare and 2 years establishment period) (1981).

Cost elements	Traditional	Improved cultivar
Labour inputs (man-days)		
land preparation (by tractor)	–	–
digging plant holes	80	200
application fertilizers & manure	40	100
planting	15	40
weeding/mulching	240	160
fertilizing	20	20
pest control	60	20
irrigation	60	40
total	515	580
Subtotal (KShs 1000)	6.0	6.8
Material inputs (KShs 1000)		
seedlings	0.6	8.0
fertilizers (at planting)	0.4	0.8
fertilizers (top dressings)	1.5	2.6
manure	1.0	4.0
chemicals (Aldrin)	2.0	0.6
mulch	1.8	1.8
tractor operations (landpreparation)	1.0	1.0
Subtotal (KShs 1000)	8.3	18.8
Total establishment costs (KShs 1000)	14.3	25.6

Source: van Santen (1982)

In Table 6.10, data on annual costs of production are given, for three different smallholder areas and for two types of estates. The information is derived from different sources and updated to represent the situation in 1981/82. Coffee production in Kenya is labour-intensive, both on estates as well as on small farms. The smallholder coffee plots (on average only 0.2 ha) can never-the-less usually be cultivated by using family labour only. The estates employ permanent labourers for all regular cultivation activities and additionally seasonal labourers during the harvesting period (e.g. the 110 ha irrigated coffee estate did employ about 80 permanent labourers).

In relatively high rainfall areas (UM1-zone) weeding requires relatively much attention, whereas in low rainfall areas (UM3-zone) mulching is one of the major activities. Weed control is carried out either by cultural methods or by using herbicides (1 l/ha Paraquat-foliage applied or 1.25 l/ha Diuron-soil acting). Estates used in the past tractor mounted cultivators, and now use chemical weed control methods. Smallholders generally apply digging with a forked hoe (jembe). When the soils are too wet, weed slashing is carried out. When coffee is grown under shade (bananas, etc.), weeding is less important. Mulching is not only a method of preserving soil moisture, but also an effective way of weed control and it contributes to soil conservation. Unfortunately it not only requires much labour, but in the densely populated coffee areas, where fallow land is scarce, some land has to be used for the production of mulch material (napier grass mainly). On the estates much time is spent on pruning (and rotational stumping), to achieve regular cropping and to facilitate picking and other operations.

Recommendations on fertilizer use are as follows: nitrogen in 3 to 4 applications from about 80 kg N for crops of less than 1000 kg per hectare to over 200 kg N for good crops (1500-2000 kg/ha); phosphate: about 50 kg/ha P_2O_5 during establishment and 10 kg/ha during the productive period. Potassic fertilizers recommendations depend on the acidity of the soil. Often the compound fertilizer 20 : 20 : 0 is applied. Nitrogen is also supplied through the application of farm yard manure. The per unit nutrient costs for nitrogen and phosphate fertilizers were in 1982 about Kshs 11 and 10 per kg, respectively.

The most serious coffee disease in Kenya is the coffee berry disease, which affects flowers, berries and sometimes leaves. Trees are not killed by this disease, but crop losses can be considerable, particularly during wet years (1967 and 1968). The Kenyan coffee breeding programme of the 1970s has been focused on the development of resistant cultivars (van der Vossen, 1981). Effective control of the disease can be achieved by the timely application of fungicides. The most commonly used varieties SL28 and SL34 are also susceptible to leaf-rust (*Hemileia vastatrix*) which also requires spraying with fungicides. A product such as Delan is effective against both diseases at a rate of 3.3 kg/ha (about Kshs 180 per kilogram).

The costs of spraying are usually too high for small farmers and many of them do not spray regularly. Estates on the other hand were already used to spray fun-

Table 6.10. Production costs of green coffee (per hectare and per kilogram) on smallholdings and estates under various ecological circumstances (1981/82).

	Smallholdings (coffee zone)			Estates	
	UM2	UM1	UM3	non-irrigated	irrigated
Yield[1] (kg/ha)	700	600	400	950	1250
Plant density (plant/ha)	1200	1300	1300	1300	1700
Area in coffee (ha)	0.3	0.2	0.15	90	110
Wage rate (KShs/man-day)	11.70	11.70	11.70	11.70	11.70
Annuity of establishment costs (KShs 1000) (10 %, 20 years)	1.7	1.7	1.7	3.1	3.6
Annual costs					
Labour inputs (man-days)					
weeding	19	50	20	90	24
fertilizing	10	10	6	20	17
disease control	4	30	4	12	34
mulching & soil conservation	22	20	40	21	10
pruning	10	20	12	70	40
irrigation	–	–	–	–	56
harvesting	147	170	80	117	172
processing	–	–	62	50	55
other (transport)	–	30	17	16	13
total	212	330	185	396	421
Subtotal of costs (KShs 1000)	2.5	3.8	2.2	4.6	4.9
Material inputs (KShs 1000)					
fertilizers & herbicides	0.5	0.4	0.5	2.4	3.3
insecticides & fungicides	1.2	0.7	0.6	2.9	2.6
processing & irrigation	–	–	–	0.8	1.2
tractor	0.5	–	0.2	0.7	1.1
other	0.7	0.2	–	0.3	0.2
Subtotal	2.9	1.4	1.4	7.1	8.4
Overhead costs (KShs 1000)					
depreciation & interest equipment	0.4	0.2	0.3	1.0	1.3
management fees & staff salaries	–	–	–	1.0	2.5
administration	–	–	–	0.5	0.4
transport and infrastructure	0.2	0.2	0.2	0.5	0.7
other (interest)	0.4	0.2	0.2	1.5	1.8
Subtotal	1.0	0.6	0.7	4.5	6.7
Total costs per ha (KShs 1000)	8.1	7.5	6.0	19.3	23.6
Costs per kilogram (KShs)	11.5	12.5	15.0	20.3	18.9
(US$)	1.21	1.32	1.58	2.14	1.99

1. Converted into green coffee equivalent, and rounded off.
2. Drying of Mbuni.

Sources: Updated information from Ruthenberg (1980) (UM2 & non-irrigated estates), Jaetzold & Schmidt (1983) (UM1), Schall & Schmidt (1976) (UM3), Magogoni estate records (pers. communication 1974) (irrigated estates)

gicides once or twice a year to retard the dropping of the leaves, which resulted in higher yields ('tonic spraying').

Harvesting is always carried out through selective picking of ripe berries. The main harvesting period is from September to December, but in the areas east of the Rift Valley there is also a minor harvesting period ('fly crop') from June to August. In some areas farmers are harvesting large amounts of unripe or overripe berries, which they dry and sell as 'mbuni coffee'. Other important cost components for estates are the depreciation and interest on buildings and equipment, the management fees, staff salaries and administrative costs.

Table 6.10 shows that the costs per kilogram green coffee (equivalent) are generally higher for estates than for smallholdings, despite the higher yields, but the costs for estates also include the processing into parchment coffee, for smallholders carried out by their cooperatives.

Table 6.11 gives an impression of the distribution of labour requirements for coffee in one particular high rainfall area, west of the Rift Valley (Kisii District). Compared with those for tea the labour requirements for coffee are heavily concentrated in a four month period. But this peak period does not coincide with cultivation period of the main maize crop (February to May) nor with the pruning period of tea.

Table 6.11. Monthly distribution of labour requirements for the cultivation of 1 ha of respectively coffee, (5th year onwards, yield of 630 kg clean coffee), hybrid maize (yield of 3000 kg per crop) and tea (4th year onwards, yield of 650 kg made tea) (1979).

	Month												
	J	F	M	A	M	J	J	A	S	O	N	D	total
Annal rainfall (Kisii)	70	89	176	265	220	146	111	151	171	138	141	107	1784
Coffee													
weeding & mulching	–	–	15	–	15	–	15	–	–	–	15	–	60
spraying & fertilizing	–	4	7	4	7	–	–	4	4	4	4	–	38
pruning	20	–	–	–	–	–	–	–	–	–	–	–	20
harvesting & transport	–	–	–	–	–	–	12	31	34	92	43	–	212
total	20	4	22	4	22	–	27	35	38	96	62	–	330
Hybrid maize													
two crops, harvested in February and August	–	44	21	27	18	–	–	44	21	27	18	–	220
Tea													
one third pruned in March	15	12	30	10	13	15	13	15	16	17	17	17	190

Source: Ministry of Agriculture (1979)

6.4.3 *Yields and returns*

National average yields of green coffee per hectare have fluctuated in the period 1971 to 1982 from 667 kg in 1981/82 and to 1199 in 1976/77. Average yields of smallholders were for some years around 730 kg/ha but dropped in 1981/82 to 540 kg/ha. Average yields for estates are in a normal year around 1100 kg/ha. With the present varieties and cultivation practices estates could under irrigation reach average yields of 1500 kg/ha. And smallholders should be able to obtain average yields of 1000 kg/ha, but the lack of proper spraying and fertilization and a large proportion of neglected holdings reduce average yields considerably.

The pricing of coffee will be discussed in Section 6.6.3 which gives the breakdown of the export unit value into the respective marketing costs and margins, and the gross and net payment to producers. The overall average gross payments to all producers were in 1980/81 and 1981/82 Kshs 21.12 and Kshs 27.64 per kilogram green coffee, respectively. For estates only few deductions have to be made to arrive at the net receipts of Kshs 19.74 and Kshs 25.90 per kilogram in 1980/81 an 1981/82, respectively. The smallholders receive their final payment through their cooperative societies, often with considerable delay, since the funds are channelled from the Coffee Board through the Cooperative Bank, the unions and the societies. The deductions include the same as those for estates, but also the processing and overhead costs of the societies, the union cess and other costs of marketing and administration. The average final payment to cooperative members for the crop years 1980/81 and 1981/82 were estimated at Kshs 15.20 and Kshs 19.95 per kilogram, respectively. The 1981/82 net payments to both estates and smallholders evidently covered the production costs incurred during that crop year, but if the prices would still have been at the 1980/81 level, these would have been close to the per unit production costs. The average producer prices conceal the different payments made according to quality. Average prices for grades 1-5 were in 1980/81 and 1981/82 KShs 25.70 and KShs 33.32 per kg, respectively.

6.5 Processing

About 90 % of the total coffee production in Kenya is fully washed arabica coffee, the remainder consisting of so-called mbuni coffee (dried cherries). Whereas the estates each have their own coffee factories, the smallholders' coffee is processed at some 700 factories (pulping installations, with an average 3 discs) owned by the 150 coffee cooperatives. The optimum level of operation of a two-disc cooperative coffee factory was found to be around 65 t of clean coffee (Kariuki, 1977), or the average production from about 90 ha. Such a factory has to be located next to a river for its washing facilities, and requires an area of about 2.5 ha for the drying tables. The present capacity of pulping facilities is just about sufficient for an average production year, but there is generally a lack of space for drying tables, in particular during years of peak production. The Coffee Research

Foundation is looking into the economics of introducing mechanical coffee dryers in the cooperative sector.

A two-disc pulping installation (McKinnon-type), including a rotary cherry feeder, a coffee pulper, a mini coffee pregrader and a Lister St 1 Diesel engine, costed in 1983 about KShs 120 000 (ex-godown Nairobi). A four-disc pulping installation, additionally including a repasser and a high pressure centrifugal water pump costed about twice as much. To these costs should be added the costs of transport and installation, the construction of buildings and stores, concrete fermenting tanks, channels, drying tables, etc., and the acquisition and preparing the terrain. The cost of depreciation, interest and maintenance of processing facilities of an average factory operating at capacity, amounted in 1981 to about KShs 1 per kilogram of clean coffee. A cooperative with 4 factories and a turnover of 250 t clean coffee has some 50 permanent staff and 150 to 250 casual labourers during the harvestig season. This represented in 1981 a cost of approximately KShs 1.50 per kilogram of clean coffee.

Most cooperative coffee factories do not have the necessary recirculation equipment, and therefore do not comply with the 1976 Water Act. Coffee pulp constitutes another pollution problem, although it is often distributed among farmers for use as fertilizers. The quality of management in the society factories has not always been satisfactory. At the Cooperative College intensive courses are held for coffee factory managers. Unfortunately the managers usually have little influence on the over-all management of the cooperative and on decisions regarding the purchase of equipment. In some areas smallholders with a sizeable area of coffee got together to build an own private coffee factory. They can obtain a licence if certain conditions are fulfilled.

The District Cooperative Unions usually undertake the transport of the pulped and dried parchment by lorry to the nearest rail head or directly to the KPCU (Kenya Planters Cooperative Union) in Nairobi, which undertakes the hulling, on behalf of the Coffee Board. The KPCU is an old cooperative formed in 1937 by large scale coffee growers around Thika. The hulling factory is able to handle up to 6 tonnes per hour. For sending in parchment bookings have to be made in advance, since the freshly processed beans need sufficient storage space and adequate handling. Upon arrival the producer's coffee is classified by sampling and tested for appearance and liquor, and on the basis of them an initial payment is made. Each producer's coffee is held separately for about ten days in order that an appeal can be made against classification.

6.6 Marketing

6.6.1 Marketing channels and functions

The marketing of coffee in Kenya is straightforward, as has been shown already in Figure 6.5. Smallholders bring their cherry coffee, immediately after harvesting, to the cooperative factory where the coffee is pulped, fermented, washed and dried. After that the cooperative societies and unions send the parchment coffee to the curing mills of the KPCU (Kenya Planters Cooperative Union) in Nairobi. The estates transport their parchment coffee directly to the KPCU. The Coffee Board of Kenya has a monopoly over the purchase of the coffee crop, and buys the coffee from the KPCU, once it is hulled, graded and classified. It then bags and stores the lots until they are auctioned on its behalf by Kenya Coffee Auctions Ltd., a firm of brokers. At the weekly auctions the coffee is sold to many independent coffee dealers, members of the Mild Coffee Trade Association of Eastern Africa.

The main marketing functions for coffee in Kenya are: assembling for processing in the pulping factories; transport and storage; grading, classification and bagging; internal pricing and taxation; selling through auctions; warehousing, export and domestic distribution. The next Section 6.6.2 deals with the important role played by the cooperative sector in the assembling stage, and Section 6.6.4 covers the final marketing stages.

Grading and classification take place at the KPCU-factory in Nairobi. The coffee passes first air separators (catadors), in which the beans are fed into a rising current of air, and subsequently gravity separators, which consist of a vibrating Table with a porous woven wire cover. Unhulled cherries or parchment, withered beans, shells and dirt are removed during these processes. The next step is the electronic colour sorting, meant to remove beans with visible defects, such as stinkers, pale or black beans and insect damaged beans. The latest KPCU-storage and grading complex with a capacity of 20 000 t clean coffee, required an investment in excess of KShs 20 million (almost US$3 million). In type and size of bean, nine different grades are distinguished and on the basis of cup quality 11 classes (10 for washed coffee and 1 for mbuni coffee) are taken into account. The major grade is usually the mediumsized normal bean (AB, screensize 15 & 16) with 40-50 % of all coffee. The major quality classes are classes 4 to 6 with about 60 to 70 % of all coffee. Whereas at least 20 % of the coffee delivered by cooperatives is top quality (classes 1 to 3), only about 3 % of estate coffee falls normally in that category. The cooperative sector, on the other hand, now delivers not less than 15 % mbuni coffee against only 5 % for estates. The districts which in recent years produced the highest-quality coffee were Kirinyaga and Nyeri Districts. The bagging is done at the KPCU and CBK-premises, stores of which are linked up by pipelines. Only pure sisal bags are used, locally produced by the east African Bag and Cordage Company.

The Government of Kenya interferes little in the export and pricing of coffee, and the taxation of the coffee sector is insignificant in comparison with that in most other coffee-exporting countries. In the period 1963 to 1973 an export tax was imposed, which originally represented about 5 % of the export unit value. In 1977 a progressive ad valorem export tax was introduced, the final rate of which relates to the prices realized on the auctions. For a certain price level the first

KShs 20 per kilogram are free, the next KShs 10 per kilogram are charged at a rate of 10 %, the next KShs 10 per kilogram at a rate of 12.5 %, etc. up to the maximum rate of 25 %. For example, at a price per kilogram of Kshs 42, the export duty would be 6 %, and at the fantastic high price of KShs 80 per kilogram, the export duty would still be only about 12 %. This export tax reduces producer prices only slightly and has hardly any stabilizing effect on the internal prices.

Warehousing of coffee is mainly undertaken by the Coffee Board, the KPCU, and some exporters, which together own about 30 warehouses with a total capacity of about 110 000 t. Because of the quota, imposed by the ICO, on exports to member countries, a large proportion of storage capacity was used in recent years. In Section 6.6.4 the export and domestic distribution of coffee is discussed.

6.6.2 The role of coffee cooperatives

Since most of Kenya's population lives in the rural areas, economic progress much depends on the development of these areas and Kenya has realized that cooperative organizations can constitute a powerful tool for improving the productivity and uplifting the living conditions of the individual peasant. Kenya did obtain its first Cooperative Societies Ordinance in 1931, but only after the 1946 revision, the formation of cooperatives among Africans was allowed. The Mau Mau rebellion and the accelerated land consolidation activities resulting from it, have stimulated the cooperative development and in particular in coffee growing areas, such as Kiambu, Kisii and Nyeri. Many local coffee pulperies were set up, which could only be run economically through the establishment of coffee marketing cooperatives. Hence membership of cooperative societies became compulsory for small scale coffee farmers. Many agricultural cooperatives were formed in the period around independence (1963), but not all of these remained active and since 1970 the emphasis has been more on strengthening existing cooperatives than on the further expansion (Ministery of Cooperative Development, 1977).

Before a separate Ministry was established, solely responsible for cooperative development, the Cooperative Department had been attached to the Ministry of Agriculture, and to several other Ministries. Compared to the Ministry of Agriculture it remained relatively understaffed, but received valuable support from the Nordic Project for Cooperative Assistance to Kenya. In 1981 there were about 50 cooperative unions and 2500 cooperative societies (half of which agricultural) with a total membership of 1.6 million. So more than 50 % of all households in Kenya are associated with the cooperative movement. Table 6.12 shows that coffee societies are the major type of cooperatives in Kenya, as far as number of societies, membership and sales of produce are concerned. The total sales value (gross income) of coffee cooperatives reached in 1981 about KShs 1300 million or about 60 % of the total sales by agricultural cooperatives. Data from the national cooperative development plan 1976 to 1980 reveal that in 1974 coffee cooperatives owned on the average assets valued at about KShs 2 million, which they

Table 6.12. Role of cooperatives in the agricultural marketing system (1981).

Cooperative	Number of societies	Number of members (× 1000)	Number of members per cooperative	Sales of produce (KShs × 10^6)	Market share (%)	Sales per member (KShs)
Coffee	169	513	3000	1322	64	2577
Cotton	40	120	3000	78	84	650
Pyrethrum	54	76	1400	144	60	1895
Sugar-cane	64	30	500	38	6	1267
Dairy	103	80	800	274	54	3425
Other	886	208	250	.	.	.
Total/average	1316	1027	780	1856	54	1807[1]

1. Farmers can be member of more than 1 cooperative (e.g. coffee and dairy).
Sources: Central Bureau of Statistics (1983), Schluter (1984)

had on average 3 employees per 100 members, and that they were able to pay their members almost 80 % of the sales value. Besides marketing and preliminary processing they render a variety of services to their members, e.g. savings, credit, farm input supply and cooperative education.

The coffee societies in Nyanza (Kisii District) and Western Provinces own usually 1 to 3 factories that serve about 500 to 2000 members, who each bring in about 60 to 100 kg coffee (green equivalent). But most coffee cooperatives in Central and Eastern Provinces operate not less than 5 to 8 factories, for an average of about 2500 to 4000 members, who each supply their cooperative with about 150 to 250 kg coffee. The 1979 issue of the Ministry of Agriculture's publication on 'Yields, costs and prices', shows pay out ratio's per district for the crop year 1975/76 fluctuating between 62 % and 85 %.

6.6.3 *Marketing margins and costs*

In the previous sections the marketing channels and agencies have been discussed, and information has been given about the prices, margins and operational costs during the respective marketing stages. In Table 6.13 a break down is given of the export unit value, obtained by the Coffee Board of Kenya in the crop years 1980/81 and 1981/82, into the different margins, taxes, costs and payments to estates, cooperatives and cooperative members. Table 6.13 shows that estates and cooperatives received in those years about 87.5 % of the total sales value, after deducting milling charges, county council cess, etc. The cooperative members received some 77 % of this sum or approximately 67.5 % of the tota sales value. The Government received only about 8 % of the sales value in the form of export duties and the respective cesses. This is low compared to the situation in many other coffee-exporting countries.

Table 6.13. Prices, marketing costs and margins[1] in KShs and US$ per kg (green coffee equivalents) (1980/81 and 1981/82).

	1980/81		1981/82		
Total sales to Coffee Board of Kenya (t)	97 717		87 436		
	(KShs/kg)	(%)	(KShs/kg)	(%)	(US$/kg)
Value sold and unsold coffee (at CBK)	22.25		29.34		
Add. income CBK (sweepings, interest)	0.24		0.34		
Average sales value (CBK)	22.49	100.0	29.68	100.0	2.83
Board expenses					
direct marketing costs	0.31	1.4	0.42	1.4	
overheads	0.09	0.4	0.09	0.3	
levies[2]	0.22	1.0	0.30	1.0	
Total	0.62	2.8	0.81	2.7	
Export duty	0.75	3.3	1.23	4.2	
Average pool payments to estates and cooperatives	21.12	93.9	27.64	93.1	2.63
Deductions					
agency fees & milling charges	0.50	2.2	0.60	2.0	
County Council cess	0.63	2.8	0.83	2.8	
average transportation costs[3]	0.25	1.1	0.31	1.0	
Total	1.38	6.1	1.75	5.8	
Average (net) receipts by estates and cooperatives	19.74	87.8	25.90	87.3	2.47
Deductions by cooperatives					
processing, marketing and adm. costs	3.74	16.6	4.90	16.5	
interest on delayed part of payment	0.80	3.6	1.05	3.5	
Total	4.54	20.2	5.95	20.0	
Average final payment to coop. members	15.20	67.6	19.95	67.2	1.90

1. Between producers and Coffee Board; excluding auction-costs and margins to exporters.
2. Levies cover largely non-marketing expenditure, such as office-expenses, publicity and research-funds.
3. By railroad and private transporters.
Source: Coffee Board of Kenya (1982)

6.6.4 *Export and domestic consumption*

The local market absorbs, with about 3000 tonnes per year only 3 % of total production. The average coffee consumption per person is only about 0.2 kg per year. For local sales two different blends are prepared by the Coffee Board, respectively named: Kenya Coffee House Blend, mainly consisting of class 3 coffee, and Kahawa no. 1, consisting of class 6 and mbuni coffee. The sales value of these blends, roasted by about 10 local firms, or by the CBU itself, was in 1984 about

KShs 20 and KShs 13 per kilogram. There are no instant coffee factories in Kenya.

As a member of the International Coffee Organization (ICO) Kenya is subject to export quotas. Due to high prices, no quotas were imposed during the period 1976 to 1979, but the ICO reintroduced quotas in 1980 and Kenya's final quota for the 1980/81 season amounted to only 67.2 t. The production in that year reached a peak of about 100 000 t, and therefore stocks increased rapidly. The quotas for 1981/82 and 1982/83 stood at 78 000 t, still more than 10 000 t below the annual production less domestic consumption. Considerable amounts had to be sold to non-quota outlets at lower prices. Of the total of sixty exporters, 5 firms ship annually more than 5000 t, whereas about 20 exporters ship between 1000 and 5000

Table 6.14. National production, domestic consumption, exports and stocks (in 1000 t) (1973-1982).

Year	Opening stock	Production	Consumption	Available for export	Export		
					total	ICO members	non-members
1973	16.6	74.6	1.2	90.0	73.4	69.5	3.9
1974	16.6	70.3	1.2	85.7	65.3	60.6	4.7
1975	20.4	73.5	1.3	92.6	80.6	78.7	1.9
1976	12.0	101.9	1.2	112.7	85.7	81.8	3.9
1977	27.0	81.4	2.6	105.8	83.4	77.7	5.7
1978	22.4	73.9	2.4	93.9	73.9	71.4	2.5
1979	20.0	99.1	3.0	116.1	82.0	79.0	3.0
1980	34.1	102.9	4.0	133.0	72.3	66.6	5.7
1981	60.7	88.2	2.2	146.7	101.9	78.1	23.8
1982	44.8	95.2	2.3	137.7	90.7	78.7	12.0

Source: ICO (1977-1983)

Table 6.15. Proportional distribution (%) of value of coffee exports to major importing countries in three years.

	1977	1979	1981
West Germany	37	47	46
Netherlands	21	9	8
United States	8	3	9
United Kingdom	4	8	7
Finland	4	6	6
Sweden	5	6	5
Italy	3	7	4
Jordan	–	–	3
others	18	14	12
Average export price (KShs/kg)	43.33	28.63	26.50

Source: Annual Trade Statistics

t. The principal market area for Kenyan coffee is Europe, with West Germany alone accounting in 1981 for 40 % of the volume and 46 % of the value of exports.

The higher qualities are generally sold to West Germany and to Finland, and the lower qualities to the United States and to the non-quota markets. The average price obtained for coffee sold to West Germany in 1981 was KShs 30 per kilogram, against an average price of Kshs 20 for sales to the United States.

6.7 Supporting services

6.7.1 Research

A considerable amount of agricultural research has been undertaken in Kenya, but until 1950 these were mainly related to the estate sector. Since the 1940s research on coffee was undertaken by the Department of Agriculture, which established in 1949 a main research station near Ruiru, and there after substations in Kisii, Meru and Koru. The management of these centres was in 1964 taken over by the Coffee Research Foundation, which is entirely financed by the Coffee Industry. Its main station is considered as one of the most advanced coffee research stations in the world. Important results have been obtained in the field of pest and disease control, fertilization, irrigation, weeding and mulching and in the field of processing. In the past ten years emphasis has been gradually changed into the direction of:

- breeding aimed at resistance to the major diseases, to avoid the expensive chemical control
- the development of compact plant types with higher planting densities and higher yields per hectare
- agro-economic research activities
- appropriate processing technology
- intensive extension activities.

In the early years of coffee production in Kenya, productive strains were developed, all selected from the original Bourbon-variety, but most of these types, appeared to be susceptible to both coffee berry disease and to leaf-rust. In the early 1970s several collections of coffee varieties were obtained from other countries, and these were tested on their resistance to these diseases and on other factors. The first new compact varieties (derived from Catimor) which combine resistance to CBD and leaf-rust with high production of good-quality coffee are expected to be ready for release to the coffee growers by 1986/87. For the large scale propagation of this planting material, hybrid seed will be produced, which is preferred above clonal (vegetative) propagation.

The introduction of the disease resistant compact varieties is economically important. It will not only reduce the import requirements of chemicals, but because of its higher yields per hectare, the total area under coffee can be brought back considerably, althus making available large areas for the production of food crops and other products. The Coffee Research Foundation established in 1979 an Economics Section, to obtain better insight into the economics of coffee production within the smallholder and the estate sector, and to get a better understanding of practices and constraints operating within the smallholder sector. The Section has been engaged in some major surveys on smallholder farms, cooperative factories and estate management.

6.7.2 *Credit and input supply*

Agricultural credit in Kenya is made available by some institutions, among which the commercial banks and AFC (Agricultural Finance Cooperation) mainly provide loans to large farmers and the CBK (Cooperative Bank of Kenya) supplies credit to small farmers through the cooperative unions and societies. Estates and medium-sized farms can obtain coffee loans from the AFC for working capital (1 year) and for processing or irrigation equipment (5 to 10 year period) at interest rates of 13 % and 23 %, respectively.

Before 1965 small farmers were virtually debarred from obtaining institutional credit, but in that year the Cooperative Production Credit Scheme (CPCS) was introduced, two years later linked up with the Cooperative Savings Scheme (CSS). The CPCS loans were mainly short term credits repayable after 18 months, and restricted to a limited number of crop and livestock enterprises (coffee, pyrethrum, sugar-cane and dairy) and hence to high potential areas). The number of participants was in 1982 close to half a million and the total loansum was about Kshs 250 million. In 1975 respectively 1976 the Smallholder Production Services Credit Programme (SPSCP) and the Integrated Agricultural Development Programm (IADP) were launched to provide production and marketing services to smallholders in particular for their food crops production. These programmes covered regions excluded from former schemes, and provided package-loans for food crops and selected cashcrops (coffee, cotton, sunflower). The repayment was tied to the marketing of cashcrops. By the end of 1982, when the SPSCP scheme was terminated and the IADP scheme had entered its second phase, these programmes had disbursed in total a sum of about KShs 75 million to about 45 000 loaners. The interest rates were in 1982 8 % and 9 %, respectively. Under these programmes, which included some grants, cooperatives were also provided with stores, pickups etc. The main purpose of the Farm Input Supply Scheme (FISS), also introduced in 1976, is to encourage the establishment of stores for resale activities within the cooperative system.

Within the general framework of the IADP, which is coordinated by a management unit at the headquarters of the Ministries of Agriculture (MOA) and Cooperative Development (MOCD), the Smallholder Coffee Improvement Project (SCIP) was established in 1980 for the rehabilitation of the smallholder coffee sector. The main objectives of this project are:

– to improve the quality of coffee produced through the construction of 14 new and the rehabilitation of 400 existing coffee factories

- provision of medium-term development loans to some 70 000 farmers, for the rehabilitation of their coffee plantations
- the strengthening of the respective organizations that provide supporting services.

The credit for the rehabilitation of factories and smallholder plantations is provided for periods of respectively 10 years with 1½ year grace period and 7 years, (4 years grace period), both at a 11 % interest rate. Since the robusta areas in Siaya and Busia districts were now included in SCIP, a separate robusta Rehabilitation Programme has also been established within MOA.

Most agricultural inputs are obtained by farmers through the KFA (Kenya Farmers Association Ltd) which organization operates 33 branches throughout the country, and through the cooperative apex organization, the KNFC (Kenya National Federation of Cooperatives) with its network of cooperative unions and societies. Apart form seed and seedlings, supplied through the research stations and cooperatives, the major inputs for coffee production are fertilizers and spray chemicals. Since 1975 the Government assumes responsibility for the import and distribution of fertilizers, and it also establishes ex-depot fertilizer prices for the respective agricultural zones. Although a local fertilizer plant commenced production in 1980, the imports of fertilizers amounted in 1981 and 1982 to Kshs 480 million and Kshs 320 million respectively, or not less than about 2 % of the total national imports.

6.7.3 Extension and training

Government services to the farming community are deliverd through five ministries and several production and marketing boards, and these are coordinated at the local level by District Development Committees. The Department of Agriculture (within MOA), with its 8 technical divisions has the primary responsibility for agricultural development. Coffee falls under the Industrial Crop Production Division. Agricultural extension work involves most of the technical divisions, but field operations are directed from the provincial level (Provincial Directors of Agriculture) and organized at the district level (District Agricultural Officers, DAO). Since 1976 the about 7000 extension agents act as generalists and each cover about 300 agricultural holdings. They operate according to the Training and visit extension system. The Ministry of Cooperative Development (MOCD) through its provincial and district officers (PCO's and DCO's) supervises the cooperative unions and societies and provides them with management advice. Since the introduction of SCIP, coffee working groups have been formed at the distict level, comprising the DAO's, the DCO's, the crops officers and the district coffee officers and factory technicians. There are about 25 district coffee officers, who are supervised by 5 provincial coffee officers. The latter report to the SCIP coffee Project Manager and the Head of the IADP-programme.

The training of agricultural staff is undertaken at the Faculty of Agriculture of the University of Nairobi (university degree level), at Egerton College (intermediate, diplome level) and at the Embu and Bukura Institutes of Agriculture (certificate level for junior staff). The number of trained agricultural staff is, however, still far from adequate. Cooperative officers and assistants receive several specialized courses at the Cooperative College, but because of the high turnover of staff, a large proportion of junior staff remains without sufficient training. There are also about 30 farmer training centres in the country.

6.8 Summary and economic parameters

Kenya is the second most important producer of arabica coffee in Africa. The agroclimatic and soil conditions in the high altitude zones in the central and southwestern part of the country are highly suitable for coffee, but the total extent of high potential zones is limited, and coffee therefore has to compete for land with several other crops, in particular maize. A large proportion of coffee plantations, and particularly among estates, are actually situated in rather dry areas, where supplementary irrigation and/or mulching is required.

Since 1970 the estate production has remained more or less constant, whereas the smallholder production has doubled and now constitutes about 60 % of national production. To reduce the impact of the coffee berry disease and spraying costs and to increase the yields per hectare, a new compact, disease resistant, variety will be introducted from 1986 onwards. It is expected that through this intensification and the higher yields per hectare, the area under coffee will decrease, benefiting the food crop production.

Because of the carefully undertaken wet processing, most coffee in Kenya falls under the highest quality category 'Colombian milds'. Whereas estates undertake this processing themselves, smallholders bring their cherries to the cooperative coffee factories, which take care of the pulping, washing, drying, storage and transport. All coffee is hulled, classified and bagged at a central factory of the KPCU, operating in conjunction with the Coffee Board of Kenya. Subsequentely the coffee is sold to exporters through weekly auctions.

Because of the high prices Kenyan coffee fetches on the world market, and through the relatively low level of taxation, Kenyan farmers receive high prices for their coffee, to some extent reflecting their high costs of operation. There are no clear price stabilization measures, except for the relatively low but with world prices progressively increasing export taxes. As a result, disproportional large funds did accrue to government, farmers and particularly to cooperative societies and unions during the coffee boom period (1976-1978), which were invested largely outside agriculture.

Economic parameters of the coffee sector in Kenya (1981/82).

Resources used in coffee sector	
Land under (productive) coffee	131 000 ha
as share of arable and permanent crop land	6 %
Labour involved	200 000 man-years
in production	150 000 man-years
in processing & marketing	35 000 man-years
in supporting services	15 000 man-years
as share of national work force	3 %
Material inputs for coffee production	KShs600 million*
Material inputs for coffee processing & marketing	KShs100 million*
Annual import requirements for coffee industry	US$50 million*
Income from coffee sector	
Value of coffee production	
(volume × export unit value)	KShs2 500 million
as share of agricultural production	11 %
as share of GDP	4 %
Value of coffee exports (f.o.b.)[1]	US$280 million
as share of agricultural exports	48 %
as share of exports	24 %
Share of total production value, obtained by	
Producers	75 %
Cooperatives, traders & exporters	15 %
Government and parastatal organizations	10 %
Other parameters	
Arabica (Colombian milds) production	99%
Yield (green coffee)	750 kg/ha
Production of labour	1.5 kg/man-day
Share of coffee produced for export	96 %
Ratio producer price (to smallholder; per kg green coffee): daily wage rate (KShs/man-day)	1.7

1. Exceptionally large volume exported in 1981/82.
Sources: FAO (1971-1983), IMF (1983), World Bank (1984), and own estimates

References

Casley, D.J. & Marchant T.J., 1978. Smallholder marketing in Kenya. FAO Marketing Development Project, Nairobi.

Central Bureau of Statistics, 1977. The integrated rural survey, 1974/75. Basic Report. Nairobi.

Central Bureau of Statistics, 1981. The integrated rural surveys, 1976-1979, Basic Report. Nairobi.

Central Bureau of Statistics, 1982. Agricultural census of large farms, 1979/80. Nairobi.

Central Bureau of Statistics, 1983. Economic Survey 1983. Nairobi.

Coffee Board of Kenya, 1982. Annual report, balance sheet and accounts. Nairobi.

Coffee Research Foundation, 1983. Coffee growers' hand-book. Nairobi.

Coffee Research Foundation, 1984. Progress Report. April 1983 to March 1984. Economics Section, Ruiru.

FAO, 1971-1983. Production Yearbook. FAO, Rome.

Heyer, J., Maitha, J.K. & Senga W.M., 1976. Agricultural development in Kenya. Oxford University Press, Nairobi.

ICO, 1977-1983. Quarterly Statistical Bulletin on Coffee (several issues). ICO, London.

IMF, 1983. International financial statistics. IMF, Washington.

Jaetzold, R. & Schmidt, H., 1983. Farm management handbook of Kenya; Volume II, Part A, B and C. Ministry of Agriculture, Nairobi.

Jomo Kenyatta Foundation, 1982. The Kenya primary atlas. Nairobi.

Kariuki, S.M., 1977. Economics of scale and economic efficiency among the cooperative coffee factories of Kirinyaga District, Central Kenya. Thesis Abstracts. The Australian National University.

Ministry of Agriculture, 1979. Farm management handbook; Vol I. Labour requirement/ availibility and economics of mechanisation. Nairobi.

Ministry of Cooperative Development, 1977. National Cooperative Development Plan, 1976-1980. Nairobi.

Roe, J.D.M. & Whitaker M.J., 1984. Economics of estate coffee production. Coffee Research Foundation, Economics Section. Ruiru.

Ruthenberg, H., 1980. Farming systems in the tropics. Clarendon Press, Oxford.

Santen, C.E. van, 1982. Economic benefits. In: Final evaluation report of the coffee breeding project in Kenya (1971-1981), Nairobi.

Schall, R. & Schmidt, H., 1976. Coffee enterprise records. Kitale.

Schluter, M., 1984. Constraints on Kenya's food and beverage exports. International Food Policy Research Institute, Washington.

Vossen, H.A.M. van der, 1981. De veredeling van Arabica koffie in Kenya. Landbouwkundig Tijdschrift.

Waters, A.R., 1969. The cost structure of the Kenya coffee industry. Rice University, Houston.

World Bank, 1979. Kenya smallholder coffee improvement project, Staff appraisal report. Washington.

World Bank, 1984. World development report 1984. Oxford University Press, Washington.

Currency equivalents and price indices in Kenya (1965-1983).

	Exchange rate (KShs per US$)	Consumer price index (1980 = 100)
1965-1972	7.14	–
1973	7.00	39
1974	7.14	46
1975	7.41	54
1976	8.37	61
1977	8.28	70
1978	7.73	81
1979	7.48	88
1980	7.42	100
1981	9.05	112
1982	10.92	135
1983	13.31	150

Source: IMF (1983)

7 Coffee in Rwanda

7.1 Background

Rwanda is a small land-locked country with common borders with Zaire (in the west), Burundi (south), Tanzania (east) and Uganda (north), respectively. The nearest seaport is Mombasa, at 1700 km from the capital Kigali. With its population of about 5.5 million (1982) and its total area of 26 300 km^2, it is the country with the highest population density on the African continent. The country has also one of the highest population growth rates: 3.6 % per year. As a consequence of the population pressure, large areas of marginal agricultural land have been brought into cultivation. Since most of the country consists of hilly and mountainous terrain, this has resulted in severe erosion problems.

The GNP per person reached in 1982 only US$260. The contribution by agriculture to total GDP is not less than 46 % (and by the food crop sector alone 37 %), against 22 % by industry and manufacturing and 32 % by services (World Bank, 1984). Of the total work force of about 2.8 million, not less than 91 % is working in agriculture, a mere 2 % in industry and 7 % in services. About 80 % of total exports consists of agricultural and agro-industrial products, than 63 % and 12 % consists of coffee and tea respectively (FAO, 1980-1982). Administratively the country is divided into ten provinces (prefectures) and 143 districts (communes) (Figure 7.1).

Table 7.1 shows the major crops in Rwanda. With an average annual production of 27 000 t in the period 1979 to 1981, Rwanda belongs to the minor coffee-producing countries and to the group of ICO members that are exempt from basic quotas. Within this group Rwanda, Burundi and Haiti are the major coffee-exporting countries. Rwanda produces predominantly semiwashed arabica coffee.

7.2 Ecology

Situated just south of the equator, Rwanda has a hilly and mountainous topography, with the average altitude increasing from about 1400 m in the east to about 2000 m in the west. On the basis of altitude, rainfall, soils and other factors the country can be divided into five agro-ecological zones:

1. The area along the shores of Lake Kivu (western border), at an altitude of 1400 to 1800 m, with an annual rainfall between 1400 and 1700 mm and with partly volcanic fertile soils. This area has the highest population density and is a major cof-

fee zone. Other major crops are beans, cassava, beer-bananas and sorghum.
2. The northern volcanic highlands, at an altitude of 1800 to 2800 m, and an average annual rainfall above 2000 mm. This zone has similar features as Zone 1. but it has a temperate climate and therefore such crops as pyrethrum and potatoes are grown.
3. The north-south mountain ranges ('Zaire-Nil') east of the above mentioned zones, at an altitude of about 2000 m and a 1400 to 2000 mm annual rainfall. Soils are less fertile and acid, and therefore formerly used for cattle grazing but because of the population pressure now an important tea zone, where also temperate food crops are grown.
4. The central high plateau, at an altitude of around 1700 m and an annual rainfall between 1200 and 2000 mm. This is the second major coffee zone, where the same crops are grown as in Zone 1. The climatic and soil conditions are good, but somewhat less than in the Zone 1.
5. The eastern sloping savannas, about 1500 m in altitude and with an annual rainfall of only 700 to 1000 mm, which generally is not well distributed. This zone includes the Akagera National Park. The presence of the tsetse fly not any longer hampers human settlement, but the zone is only marginally suitable for most traditional crops, including coffee.

Almost 30 % of the total area is cultivated with food crops (at an average cropping intensity of 150 %), and 2 % with export crops (coffee and tea mainly), whereas about 18 % of the total area consists of pasture, fallow and 20 % of non-arable slopes. The remaining 30 % consists of parks, lakes, rivers and forested areas. The agricultural development is mainly at the expense of the traditional grazing areas and the livestock is moved towards the non-arable slopes.

Although not exclusively, coffee is mainly grown in Zone 1 and Zone 4. On the basis of the coffee yields, five specific coffee areas (terroirs à cafe) are distinguished: areas with high and very high yields (800 kg/ha) are only found at the northern shores of Lake Kivu; reasonable and moderate yields are obtained along the southern shores of Lake Kivu and in some areas on the central plateau, and low yields (400 kg/ha green coffee) are obtained in other parts of the central plateau and in other provinces (Figure 7.2).

7.3 The coffee sector

7.3.1 History

Coffee was first established in Rwanda in 1905 with seed of Guatemalan origin. This variety, called Mibirizi, after the mission where it was first planted is still the main variety in both Rwanda and Burundi. Although well-adapted to local conditions, this variety is not high-yielding, and some better varieties were introduced from Réunion and Kenya.

From 1927 onwards, the colonial authorities actively promoted coffee by plant-

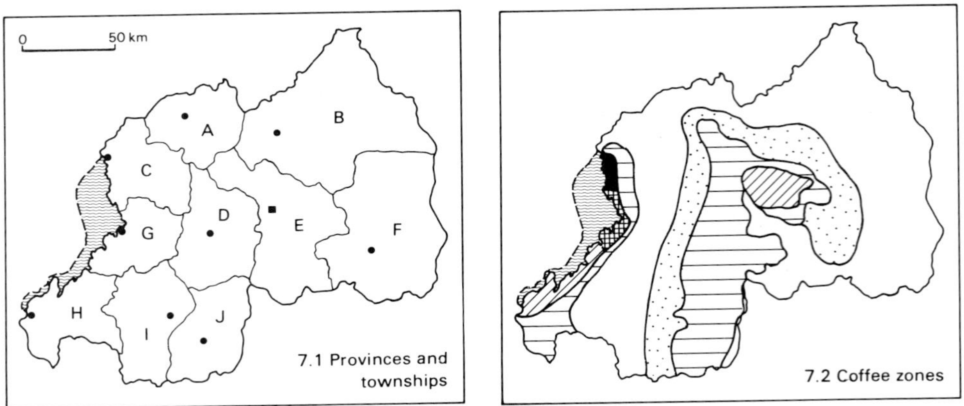

Fig. 7.1-7.2. Provinces and townships (7.1) and the production level in coffee zones (7.2) of Rwanda. After ISAR (1969).

Table 7.1. Average area harvested, yield and national production of major crops over 3-year periods.

Crops	Area (1000 ha)		Yield (kg/ha)		Production (1000 t)	
	1969/1971	1979/1981	1969/1971	1979/1981	1969/1971	1979/1981
Food crops						
banana & plantains	146	225	11342	9147	1656	2058
beans (dry)	161	216	900	815	145	176
sorghum	132	149	1068	1134	141	169
sweet potatoes	65	116	5788	7839	379	906
Oil crops						
groundnuts	8	18	869	948	7	17
soya beans	1	6	781	821	1	5
Beverages						
coffee	24	32	578	859	14	27
tea	2	6	778	1071	1	6
Other crops						
pyrethrum	2	2	700	600	1	1
quinquina	1	2	700	500	1	1

Source: FAO (1971-1983)

ing it alongside roads and tracks, in an effort to provide smallholders with a cash income. Today coffee is in fact the major source of cash for the farmers in the country. During the last two decades the area under coffee has more than doubled, from about 15 000 ha in 1960 to 32 000 ha in 1980.

7.3.2 *Recent developments*

Since the 1950s Rwanda depends for its foreign exchange heavily on the production of coffee. The production rose steadily until 1976, but stagnated in recent years. Exports often exceed domestic production by some 20 %, through illicit border trade with Zaire and Uganda. To diversify exports the production of quinine has been promoted in the major coffee areas and tea and pyrethrum in the high altitude zones. In the past ten years much attention has been given particularly to the production of tea, which increased considerably (Table 7.2).

In all provinces, considerable amounts of coffee are produced, almost exclusively on small, privately owned plots. In the provinces Cyangugu, Gitarama and Kigali the total area under coffee is relatively large, whereas in the densely populated and high potential areas of Gisenyi and Kibuye provinces remarkably high yields are obtained (Table 7.3). On average, households produce very little coffee; only 36 kg if all farms are considered and about 70 kg among the households that produce coffee. Hence the average earnings per farmer obtained from coffee are low, but more than half a million farmers do obtain some cash income from coffee.

Table 7.2. Area planted, national production and average yield of green coffee and made tea (1960-1981).

Year	Coffee[1]			Tea		
	area (1000 ha)	production (1000 t)	yield (kg/ha)	area[2] (1000 ha)	production (1000 t)	yield[2] (kg/ha)
1960	15.3	10.5	686	.	.	.
1965	18.3	9.9	538	.	.	.
1970	23.9	14.2	595	1.6	1.2	750
1975	28.8	18.0	626	5.5	4.0	730
1976	30.8	20.3	660	6.2	5.0	810
1977	31.9	15.3	471	6.9	5.4	780
1978	32.4	16.3	502	7.8	5.3	680
1979	29.1	17.5	611	8.4	5.7	680
1980	29.6	15.0	507	8.9	6.6	740
1981	31.9	19.4	608	9.2	6.8	740

1. Conversion factor green coffee/parchment is 0.74.
2. Total area includes new, not yet productive plantations.
Source: Ministère de l'Agriculture et de l'Elevage (1981)

Table 7.3. Area planted, national production and average yield of green coffee, and production and revenues per farmer, by province (1981).

Province	Area (1000 ha)	Proportion of total agric. land (%)	Production (1000 t)	Yield (kg/ha)	Farmer	
					production (kg)	revenues (FRw)
Butare	4.0	3	2.0	500	27	4420
Byumba	1.8	1	1.0	594	27	4420
Cyangugu	4.0	6	2.2	545	49	7880
Gikongoro	2.2	3	0.8	373	24	3940
Gisenyi	1.8	2	2.1	1139	65	10500
Gitarama	5.1	4	3.3	643	35	5610
Kibungo	3.9	3	2.4	618	38	6090
Kibuye	2.5	4	2.2	880	53	8600
Kigali	5.3	3	2.8	531	35	5610
Ruhengiri	1.3	1	0.6	444	20	3220
Total/Average	31.9	3	19.4	608	36	5850

Source: Ministère de l'Agriculture et de l'Elevage (1981)

7.3.3 Organization

Coffee is in Rwanda almost exclusively grown by smallholders. Their plantations consist on average of about 150 trees, which constitutes some 10 % of the average farm size of 1 ha. Some coffee plantations, in total 50 ha, have been established through the 'umuganda' community work programmes by government departments. The most well known plantation is that of the Defence Ministry near the airport. Farmers bring their cherry coffee to one of the about 1000 pulping centres, which are generally equipped with small hand pulping machines and some minor washing facilities (only little fermentation is carried out). They subsequently sell the parchment coffee to traders who bring it to one of the three hulling factories.

After the hulling and bagging the coffee is classified and stored by OCIR-Café. OCIR was after independence in 1962 created as a parastatal company under the Ministry of Agriculture and Livestock to look after all operations with regard to coffee and other industrial crops (tea, pyrethrum and others). It was reorganized in 1978 and split into OCIR-Café and OCIR-Thé, respectively. The Director of OCIR-Café is designated by the President, and supervises the operations of the three services responsible for: Administration and Finance; Production; and Commerce, respectively. The production service of OCIR-Café is among others responsible for extension services (through the 'communes'), nurseries, pest control and pulping centres, whereas the commercial service is involved in the preparation of sales contracts, classification and quality control, pricing structure, stor-

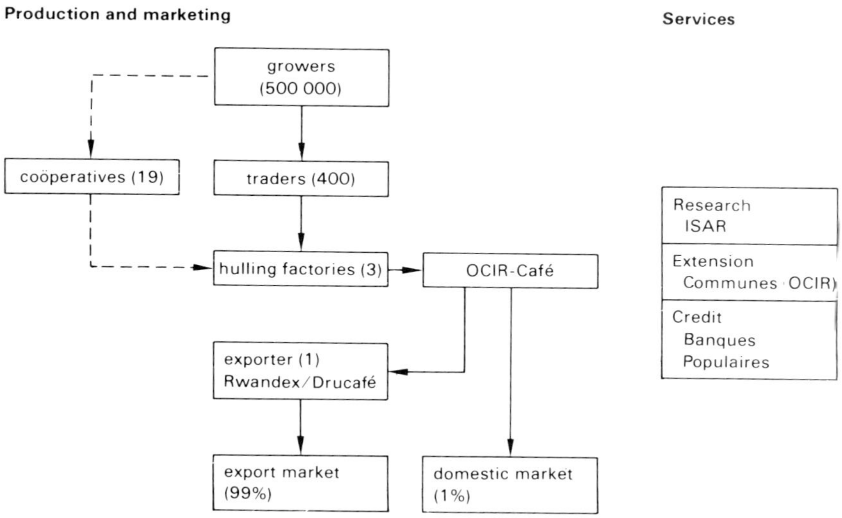

Fig. 7.3. Participants in the coffee sector and main marketing channels in Rwanda in 1982. Solid line, main channel; broken line, secondary channel.

age and transport coordination. The production service has two branches for the promotion of the coffee production and the production of other crops (mainly quinine). OCIR-Café employs about 200 professional and support staff, and about 300 labourers and it pays the salaries of around 300 coffee extension workers. The total operating budget of OCIR-Café for 1980/81 was about FRw 260 million of which FRw 75 million was spent on salaries and around FRw 60 million on insecticides (OCIR-Café, 1981). Apart from its regular budget, OCIR-Café administrates two special funds: the stabilization fund and the propaganda fund. The stabilization fund depends for its revenues on the ratio between the export price and the internal costs (Section 7.6.2) and the propaganda fund receives its income from the local sales of coffee and from part of the ICO-tax.

Coffee research is carried out by ISAR. The coffee trade is delegated to the company Rwandex, which transports the green coffee to Mombasa (by road or by air). From there it is further exported by the private company Drucafé. The Government has left the domestic marketing, the processing and the export-marketing entirely in private hands, but establishes narrow margins for these activities in the annual coffee price structure.

7.4 Production

7.4.1 Production units and farming systems

Rwandan agriculture is completely dominated by smallholder production which is oriented primarily towards food crop production for home consumption. The

Table 7.4. Proportional distribution (%) of farm size and cropping pattern for average size of farm.

Farm size (ha)		Cropping pattern	
less than 0.49	26	Pure stand: bananas	8
0.50-0.99	31	food crops	16
1.00-1.49	15	cash crops (incl. coffee)	6
1.50-1.99	9	Mixed cropping	30
2.00-2.99	12	Not cultivated or fallow	40
more than 3.00	7		

Source: Ministère de l'Agriculture et de l'Elevage (1983)

total number of holdings is just over a million and the average size is about 1.1 ha; but the median is not much more than 0.8 ha (Table 7.4). The holdings are scattered in a patchwork, usually over a hillside ('colline') and villages do not exist. The farm unit is the nuclear family, with an average of 5 people, of which about half within the age group 16 to 65 years. Whereas both men and women work in the field, the men are principally in charge of cashcrops and the cattle and the women take care of the food crops and the small stock. The men are usually also involved in some off-farm activities and both men and women are to participate in the unpaid 'umuganda' community activities.

The production practices applied are overridingly traditional; rainfed with neither mechanical nor animal traction, little use of modern material inputs and despite the steep terrain only limited application of erosion control measures (Musema-Uwimana, 1983). But food crop production is intensive, with an average cropping intensity of around 150%. Major crops at medium altitude are beans, sorghum, sweet potatoes, maize, cassava, coffee and bananas, and on higher altitudes emphasis is laid more on peas, potatoes, tea and pyrethrum. Table 7.4 shows that a major part of the cultivated land is under mixed cropping, the basic crops among which are usually bananas, beans and sorghum.

By law, all land in Rwanda belongs to the Government and uncultivated land can be used by the Government for any purpose. In practice, individual usufruct rights exist on continously cultivated land, which can be passed on to male heirs, whereas pasture is communally held. The use of land within settlement schemes ('paysannats') is, however, regulated by contract. There are about 20 of such schemes in 4 regions and a total of about 35 000 settlers produce some coffee.

Coffee farmers generally devote one little plot of between 6 to 20 are to coffee (90 to 300 trees). The use of fertilizers is limited on coffee plots, and for economic reasons is not actively promoted. Instead farmers are obliged to apply mulching, to provide nutrients, to facilitate weedcontrol, to conserve moisture and to contribute to soil conservation. Sources of mulch are among others straw and crop residues of sorghum and beans, banana-leaves and several grasses, part of which is cut from the grass-strips used for erosion control. But because of the increasing pressure on the land, and the requirements for livestock feeding, the amount of mulch material has diminished rapidly, and mulching is no longer generally applied.

The rule of keeping coffee as a pure stand is also not always respected any more, and pruning is not carried out regularly. Recent measures, such as the replacement of the traditional varieties by two better varieties (Jackson-2 and Bourbon Mayaquez-39) and the introduction of higher plant densities (from 1600 to 1750 trees/ha), have therefore not resulted in higher coffee yields. A major coffee pest in Rwanda is antestia (*Antestiopsis orbitalis*, the variegated coffee bug), against which Sumithion-3 % is applied. OCIR provides free of charge spraying equipment and insecticides to farmers, as well as other material inputs. Selective picking is applied during harvesting.

7.4.2 *Production costs, yields and returns*

Although the costs of production per unit area do not vary much between farmers in the respective areas, or only according to the state of maintenance, the yields vary by coffee zone and hence the production costs per kilogram coffee. Thanks to soil and climatic conditions along the shores of Lake Kivu, the production per tree of parchment coffee in those areas is 800 g or more, which gives a production of at least 560 g of green coffee (high potential zones). In the central highlands and surrounding areas the production per tree is not usually more than 500 g of parchment or 375 g of green coffee (moderate potential areas).

In reviewing the costs of production, it is important to realize that the size of coffee plots in Rwanda is generally only about 0.1 ha. For comparison, production costs are, however, provided on a per hectare basis. Table 7.5 gives the costs of

Table 7.5. Establishment costs (1-4 years) per hectare of coffee (1982).

Cost elements	1	2	3	4	Total
Labour inputs (man-days)					
clearing & ploughing	180	–	–	–	180
digging of holes & manuring	80	–	–	–	80
transport, planting & shade	80	–	–	–	80
mulching	220	–	–	–	220
weeding & cleaning	20	50	50	50	170
total	580	50	50	50	730
Subtotal (FRw 1000)					88
Material inputs (FRw 1000)					
plants (FRw per plant)	9	1	–	–	10
other material	2	2	2	2	8
Subtotal (FRw 1000)	11	3	2	2	18
Total establishment costs (FRw 1000)					106

Source: Personal Communications

establishment, which include general land-preparation activities, erosion control measures, digging of planting holes, manuring, mulching, the raising and planting of coffee plants and shade trees, and the annual maintenance over a three-year period. The labour inputs are considerable, but for an average farm these should be divided by ten, and are to be valued at the prevailing, relatively low, wage rate of Frw 120 per man-day. Considering a 10-20 % replanting rate, about 2000 plants per hectare are required. These establishment costs are added to the annual production costs, in the form of an annuity at 10 % over a 20-year period (Table 7.6). The establishment of 1 ha coffee requires 730 man-days, against more than 1000 man-days for tea (hilly terrain).

Table 7.6. Production costs of green coffee (per hectare and per kilogram) of two zones (1982).

	Moderate potential	High potential
Yield (kg/ha)	600	1000
Plant density (plants/ha)	1600	1800
Production per tree (gram green coffee)	375	560
Production per farm (kg)	45	90
Wage rate (FRw/man-day)	120	120
Annuity of establishment costs (FRw 1000) (10%, 20 years)	12.5	12.5
Annual costs		
Labour inputs (man-days)		
weeding & pruning	45	45
mulching & manuring	45	45
disease control	15	15
harvesting	100	160
processing	20	32
drying	18	28
transport	12	18
total	255	343
Subtotal (FRw 1000)	30.6	41.2
Material and other costs (FRw 1000)		
mulch material (opp. costs)	10.0	10.0
chemicals (Sumithon 3 %; 60 kg)	6.0	6.0
bags and baskets	1.0	1.5
depreciation of interest on drying screens	8.0	8.0
depreciation and interest on tools	6.0	8.0
Subtotal	31.0	33.5
Total costs per hectare (FRw 1000)	74.1	87.2
Costs per kilogram (FRw)	124	87
(US$)	(1.34)	(0.94)

1. Conversion factor (green) coffee/parchment = 0.74.
Sources: OCIR-Café (1983), World Bank (1983a)

In the annual production costs a distinction is made between the two major coffee areas, referred to above. Whereas the labour inputs for general cultivation practices are the same for both areas, the labour inputs for harvesting and processing are related to the yields obtained. The value of material inputs is rather high, which is largely because each farmer, whatever the size of his plot at least requires the minimum equipment, including such items as pruning knife, basket, spraying equipment, drying screen ('claie de sèchage'), etc. The total costs are about FRw 73 000 to 86 000 per hectare or about FRw 7000 to 9000 per farm, and the costs per kilogram green coffee are, according to these calculations, FRw 86 for high-potential areas and FRw 122 for areas with only moderate potential.

Usually a distinction is made in Rwanda, between five coffee zones (Section 7.2) with yields ranging from only 400 kg/ha in less suitable areas to about 1200 kg/ha in Gisenyi province along Lake Kivu. In these two extreme cases production costs are most likely about FRw 140 and FRw 70 per kilogram green coffee.

Farmers receive since 1977 for their coffee FRw 120 per kilogram parchment which is equivalent to FRw 162 per kilogram green coffee. So the return to labour for farmers in the zones with moderate potential was in 1982 around FRw 210 per man-day, and for those in the high potential zones not less than FRw 340 per man-day. Compared with other crops, the gross returns per man-day for coffee are only exceeded by bananas and cassava, whereas these crops show slightly lower gross returns per hectare. Other food crops and tea give lower figures (Ministry of Planning, 1979). The annual labour requirements for the production of tea (3 year cycle) are about 450 man-days per hectare, at a production level of 4000 kg/ha green tea (900 kg made tea), and a farmgate value of about FRw 15 per kilogram (green).

7.5 Processing

A major part of the coffee processing takes place at or more often near to the farm. The ripe cherries are pulped at the hand pulping centres, which consist of hand-driven disc pulpers and at best a small washing basin, where part of the mucilage can be removed (semiwashed coffee). No fermentation takes place. Most of these centres (1100 in total) were established before independence, usually on a per hill ('colline') basis, and they serve a hundred or several hundred farmers (the aim is one for 50 000 coffee trees).

Now these centres are, where possible, maintained by OCIR, but they are not always effective any more. In recent years some hundred new centres were created. Pulping facilities for one centre, including the installation, cost about FRw 400 000; hence annual costs of depreciation, interest and maintenance are around FRw 60 000. Already in 1956 efforts were undertaken to produce high quality, fully-washed coffee, through the establishment of two cooperatively operated coffee washing stations, but they were for several reasons not very useful. Proposals are now made for the establishment of mechanical pulpers with fermentation facilities to produce fully-washed coffee, which could possibly fetch an additional FRw 12 per kilogram green coffee. After pulping, farmers take the wet parchment coffee back home, to dry it in the sun. The high rainfall in the harvest season (March to June) unfortunately affects the drying, and makes it impossible to apply the dry processing method.

Once sufficiently dried, the parchment coffee is sold to traders, who bring it with their vehicles to one of the three hulling factories; the Rwandex's factories in Kigali or Gisenyi, or the Etiru factory at Ruhengiri. The Government has 51 % interest in Rwandex, whereas the Etiru factory is privately owned. The factories are in operation from the beginning of June to the end of September. They employ about 300 permanent staff and about 400 seasonal labourers. The factories are paid FRw 5 per kilogram green coffee for hulling and bagging.

7.6 Marketing

7.6.1 Marketing channels and functions

Before every coffee marketing season, which starts in May, the Government fixes the producer prices, the margins to traders and the respective commissions. A certain number of traders, often about 4 per commune, are licensed by the 'Prefect' of each 'Prefecture', to buy coffee from the producers. Several hundred private traders and about ten to twenty cooperatives are involved in the local coffee trade. They use small pickups and 7 to 8 tonnes lorries, respectively, for the transport of parchment to the factories and generally obtain credit from private banks. The traders in 1983/84 received a commission of 6 FRw per kilogram parchment, which is equivalent to 8.2 FRw per kilogram green coffee.

From the factories the coffee is sent to the OCIR stores in Kigali, where the coffee is classified on a sample basis. The samples undergo a whole range of tests, as green coffee, as roasted coffee and as beverage, respectively. The classification is based on the following criteria, respectively for :
- green coffee: bean size, regularity and colour
- roasted coffee: appearance (bright/dull), regularity
- beverage: acidity, body, cup quality (taste).

Based on these aspects five classes are distinguished: Superior, Standard, Ordinary, Ungraded and Triage. Coffee that has passed the fermentation stage forms a separate category: Fully washed. The cup quality of Rwandan coffee is usually excellent, but at times a defective flavour appears, caused by a pest, which becomes more pronounced when the coffee is highly roasted. Therefore the United States market, with its light roasts, is more accessible than the European (Marshall, 1983).

The coffee is transported by Rwandex to Mombasa, and afterwards exported by the private company Drucafé. In the period 1980/81 the situation in Uganda prevented regular transport by road, and coffee was then sent by air to Mombassa. Since then, coffee is transported by road as well as by air.

7.6.2 *Pricing and taxation*

During the coffee boom in 1977, the producer price was raised from FRw 65 to FRw 120 per kilogram of parchment (or from FRw 89 to FRw 164 per kilogram of green coffee equivalent). Since this considerable price increase, the producer price has not changed any more, and due to inflation the real price was in 1982 equal to that in 1976.

Despite the high payment to producers, the export unit value obtained in the period 1977 to 1979, left enough room for the absorbtion by the Government of important revenues through taxation (Table 7.7). Prior to 1979 there was not only a customs tax, at the rate of 20 % of the border value (internal costs plus transport costs to the border: Kigali-Kagitumba), but also a progressive tax on coffee ex-

Table 7.7. Marketing costs of green coffee production and taxation (in Frw 1000 per kilogram) in three years.

Cost elements	1977	1980	1983
Internal costs			
price to producer	164.4	164.4	164.4
traders' margin (collection)	13.0	14.0	8.2
factory margin (hulling, bagging)			5.0
OCIR-taxes remuneration			7.7
transport-subsidy	8.3	9.0	2.0
ICO-contribution			0.3
subtotal (a)	185.7	187.4	187.5
Government Taxes			
(PM) transport Kigali-Kagitumba (b)	0.8	0.8	1.4
customs tax (20 % of value = a + b)	37.3	37.5	37.8
progressive tax	41.6	–	–
External costs			
transport Kigali-Mombasa[1]	23.3	17.0	14.5
transit to Mombasa	.	.	1.7
insurance, financial charges	3.0	3.2	4.5
port charges (1977/80 incl. transit)	1.6	2.0	1.4
allowance for price increase	1.0	1.0	0.5
exporter commission (1.5 % f.o.b.)	3.0	3.2	3.2
subtotal	31.9	26.4	25.8
Total costs per kilogram	296.5	251.4	251.2
Export unit value f.o.b. Mombasa (FRw)	347.5	264.2	250.0
(US$)	(3.74)	(2.85)	(2.58)
Surplus/deficit	+51.0	+12.8	−1.2

1. Transport by road; transport is also carried out by air for which the following external costs were estimated for 1983: Transport: FRw 17500; Transit FRw 2450; Insurance (0.053 %): FRw 108; other items, same as above.

Source: OCIR-Café (1983), World Bank (1983a)

Table 7.8. Tea prices and distribution of earnings (1976-1980) (FRw/kg).

	1976	1977	1978	1979	1980
Export unit value (f.o.b. Mombasa)	120	211	154	173	155
Producer (ad factory) price	37	49	46	66	51
Processing and marketing costs	35	50	64	65	63
Transport costs	16	22	33	33	32
Financial & administration costs	5	7	3	1	1
Depreciation	6	8	8	8	8
OCIR-Thé margin	21	75	(8	8	8)

Source: World Bank (1983a)

ports. This tax was abolished in August 1979, and since then the following rules are applied: when the f.o.b.-Mombasa price falls below Frw 255 (US$2.75) per kilogram, the Government subsidizes the coffee; between FRW 256 and FRw 293 (US$3.16) per kilogram the excess between f.o.b. price and costs are equally distributed between the stabilization fund and the Treasury and over FRw 293 the fund receives FRw 18.80 per kilogram and the Treasury receives the rest.

These different situations are illustrated, in reverse order, in Table 7.7, for the periods 1977, 1980 and 1983, respectively. In recent years, from 1980 onwards, there was hardly any excess between the f.o.b. price and total costs and the customs duties constituted the only revenues to the Government. Whereas in the years before 1977 farmers received only about 40 % of the export value of their coffee, in the period 1977 to 1983, they received on average about 60 %. The share in the export value obtained by tea farmers is much less, as shown in Table 7.8, since tea requires higher costs of processing.

7.6.3 Export and domestic consumption

The national coffee production has fluctuated in the period 1977 to 1982 between 15 000 and 29 000 t of green coffee. Little is known about the domestic consumption, but this represents only a fraction of total production. Table 7.9 clearly illustrates that the exports of coffee include coffee smuggled into Rwanda from neighbouring countries. This last category constituted in the period 1977 to 1982 about 20 % of total exports. Traditionally most of the Rwandan coffee has been exported to the United States, but in recent years the exports to other countries increased, particularly to countries of the European Community. Surprisingly Belgium, of which country Rwanda was a protectorate and which is still one of the major trading partners of Rwanda, does not feature among the major coffee importing countries.

Table 7.9. National production, domestic consumption and exports (in 1000 t) (1977-1982).

Year	Opening stock	Production	Consumption	Export licences issued	Available for export	Export	Export-value (FRw 1000)
1977	0.3	17.0	0.3	21.3	21.6	19.1	6638
1978	2.5	21.9	0.2	26.1	28.6	17.2	4533
1979	11.4	19.8	0.1	30.6	42.0	39.3	12820
1980	2.7	15.0	0.1	30.6	33.3	22.0	5813
1981	11.3	21.1	0.1	29.1	40.4	30.0	6342
1982	10.4	29.0*	0.1	.	.	26.3	6680

Sources: OCIR-Café (1981), World Bank (1983b)

Table 7.10. Volume and proportional distribution of coffee exports to major importing countries (1972-1982).

	1972/73-1974/75		1979/80-1981/82	
	(1000 t)	(%)	(1000 t)	(%)
United States	16.5	74	15.0	54
Netherlands	0.6	3	2.8	10
West Germany	0.6	3	2.4	9
United Kingdom	0.4	2	1.9	7
Finland	–	–	1.2	4
Sweden	0.1	1	1.0	4
Algeria	.	.	0.9	3
Yugoslavia	0.5	2	0.9	3
France	1.4	6	0.3	1
others	2.1	9	1.4	5

Source: OCIR-Café (1981)

7.7 Supporting services

7.7.1 Research and extension services

Agricultural research is carried out, more or less exclusively, by ISAR (Institut des Sciences Agronomique du Rwanda), a parastatal organization under the Ministry of Agriculture and Livestock. It has at its disposal a central station, covering 675 ha, at Rubona, several other regional stations and a staff among which about fifty academic research workers.

The emphasis has been laid particularly on the improvement and expansion of major cashcrops and livestock but also on the production of improved seed for food crops, on better cultivation practices and on the organizing the 'paysannats' (settlement schemes). In the past research results have not been tested sufficiently under farm conditions, and therefore had limited impact.

One topic of discussion were the fertilizer trials, which were given less attention after 1974, when fertilizer prices increased rapidly. Pleas were then made to orient small farms further towards the use of farm-produced inputs (mulch, manure). Ideas were developed for a better crop-livestock integration on small farms, to be transformed in so-called 'fermettes' (Sauve, 1977). The question remains if many of the 'mini-farms' in Rwanda can afford to use some land for foddercrop production. The cattle population is steadily decreasing, whereas small stock is on the increase. Production systems should at any rate be both land and labour-intensive, and include high-valued products. More use could also be made of crop-by products such as coffee pulp. The new Lake Kivu-project for the improvement of coffee and food crop production, emphasizes again the need for fertilizer trials, apart from applied research in the field of disease control, regeneration of coffee trees and soil conservation measures.

Socio-economic research is the domain of INRS (Institut National de Recherche Scientifique), situated in Butare. This Institute has among others carried out studies on the nutritional situation in Rwanda. Both institutes (ISAR and INRS) have been until recently supported by Belgian aid and were linked up with institutes in Belgium.

Before 1975 the Production Service of OCIR-Café was responsible for the coffee extension agents, but in that year the responsibility was transferred to the 'communes' (districts). OCIR-Café is still paying the salaries of the about 300 extension workers. Realizing that the reduction in yields can be attributed largely to the absence of appropriate agronomic measures, such as mulching, pruning, spraying, and soil conservation, the Production Service assists the local authorities with pest control, pruning and other campaigns and also with the establishment of nurseries and with the installation and maintenance of the hand pulping machines (through so-called 'Brigades des Centres de Dépulpage'). But generally the extension agents are poorly organized and supervised, lack means of transport and are inadequately trained. Besides there is no formal link between extension services and research activities. For the Lake Kivu-project it is planned that extension workers will direct themselves to small groups of (6) farmers, which they will visit at regular times, at least once a month.

7.7.2 Credit and input supply

The 'Banques Populaires' are the only financial institutions actively involved in provision of agricultural credit. But they were generally more successful in mobilizing rural savings, to be lent outside the agricultural sector, than in identifying and using agricultural credit opportunities. On the other hand most services rendered to farmers are heavily subsidized. In 1980/81 about 8 million coffee plants were raised, (from 5 t seed) in the respective nurseries in the country, and distributed free-of-charge to farmers. For the spraying against antestia, carried out in July and December by OCIR-staff, a total of 1400 t insecticides (Sumithion) was distributed. In the same year not less than 11 000 pruning knifes and 1500 hand spraying equipment were given to extension agents, who unfortunately do not take proper care of that equipment. OCIR realizes that it would be better to provide farmers themselves with such equipment.

7.8 Summary and economic parameters

Rwanda belongs to the group of minor coffee-producing countries, which for their foreign exchange earnings depend heavily on the exports of coffee. Situated close to the equator, with the attitude ranging from about 1400 to 2000 m, a mean annual rainfall between 1000 and 2000 mm, and volcanic soil types in several

Economic parameters of the coffee sector in Rwanda (1982).

Resources used in coffee sector	
Land under (productive) coffee	30 000 ha
as share of arable and permanent crop land	3-4 %
Labour involved	45 000 man-years
in production (incl. on farm processing)	40 000 man-years
in processing & marketing	3 500 man-years
in supporting services	1 500 man-years
as share of national work force	2 %
Material inputs for coffee production (inclusive depreciation equipment)	FRw400 million*
Material inputs for processing & marketing	FRw500 million*
Annual import requirements for coffee industry	US$6 million*
Income from coffee sector	
Value of coffee production	
(volume × export unit value)	FRw7 000 million
as share of agricultural production	12 %
as share of GDP	5 %
Value of coffee exports (f.o.b. Mombasa)	US$72 million
as share of agricultural exports	78 %
as share of exports	63 %
Share of total production value, obtained by	
Producers	63 %
Processing industries, private traders & exporters	17 %
Government	20 %
Other parameters	
Arabica (semi washed) production	98 %
Yield (green coffee)	700 kg/ha
Production of labour involved (green coffee)	1.5 kg/man-day
Share of coffee produced for export	99 %
Ratio producer price (per kg green coffee):	
daily wage rate (FRw/man-day)	1.35

Sources: FAO (1971-1983), IMF (1983), World Bank (1984), and own estimates.

areas, the country offers good growing conditions for arabica coffee. Since the rainfall distribution is not adequate in several areas, mulching is generally applied. In fact not only for moisture conservation, but also for fertilization and soil conservation purposes.

Rwanda is the most densely populated country in Africa and the average farm size is only about 1 ha. A large proportion of farmers do cultivate coffee, on plots of around 0.1 ha. Cultivation practices are labour-intensive and reasonable yields are obtained (700 kg/ha). The farmers undertake the processing at small pulping centres in their neighbourhood, where because of little fermentation so-called semiwashed coffee is obtained.

Subsequently the coffee is sold at quaranteed prices to traders who transport it to one of the three hulling factories in the country. The coffee is classified in stores of the national coffee organization OCIR, and all coffee is exported by one company. Rwanda is a land locked country and the coffee has first to be transported to Mombasa, the main port of Kenya, either by road or by air. More than half of Rwandan coffee is exported to the United States.

References

FAO, 1971-1983. Production Yearbook. FAO, Rome.
FAO, 1980-1983. Trade Yearbook. FAO, Rome.
IMF, 1983. International financial statistics. IMF, Washington.
ISAR, 1969. Terroirs à café. Rubona.
Marshall, C.F., 1983. The world coffee trade. Woodhead-Faulkner, Cambridge.
Ministère de l'Agriculture et de l'Elevage, 1981. Rapport annuel 1981. Kigali.
Ministère de l'Agriculture et de l'Elevage, 1983. Enquête nationale de l'Agriculture. Resultats préliminaires de la phase pilote. Kigali.
Musema-Uwimana, A., 1983. La conservation des terres au Rwanda. Bulletin Agricole du Rwanda, April 1983. Ministère de l'Agriculture et de l'Elevage. OCIR-Café, Kigali.
OCIR-Café, 1981. Rapport annuel 1981. Kigali.
OCIR-Café, 1983. Calcul des prix du café, 1983/84. Kigali.
Sauve, G., 1977. Description de la situation agricole en Rwanda. Groupe scolaire de Butare, Butare.
World Bank, 1983a. Rwanda agricultural strategy review. Washington.
World Bank, 1983b. Rwanda economic memorandum – Recent economic and sectoral developments and current issues. Washington.
World Bank, 1984. World development report. Oxford University Press, Washington.

Currency equivalents and price indices in Rwanda (1965-1983).

	Exchange rate (FRw per US$)	Consumer price index (1980 = 100)
1965	50.0	.
1970	100.0	30
1971	99.7	30
1972	92.1	31
1973	83.9	34
1974	92.8	45
1975	92.8	59
1976	92.8	63
1977	92.8	72
1978	92.8	81
1979	92.8	93
1980	92.8	100
1981	92.8	107
1982	92.8	120
1983	94.3	128

Source: IMF (1983)

8 Coffee in Cameroon

8.1 Background

Cameroon has a total area of 475 000 km^2 and a population of 9.3 million (1982). The country is situated more or less in the centre of the African continent, and encompasses many typical features of Africa. Ecologically one finds a tropical rain forest climate in the south-west and semidesert conditions in the north. Historically there were several important kingdoms and chieftainships that fought each other and were broken up by the colonial powers. And the great ethnic diversity has stimulated the use of several vehicular languages (Pidgin, Foulfonde, Arabic) next to the two official languages: French and English.

Administratively the country was until recently subdivided into seven provinces, of which one (the Northern province) was in 1983 split up into three provinces. The provinces are subdivided into regional divisions, departments and districts.

The GNP per person reached US$890 in 1982. The GDP was in 1982 distributed as follows: agriculture 27 %, industry 31 % and services 42 %. Of the total work force of 4.6 million, about 83 % was in 1981 still working in agriculture against 7 % in industry and 10 % in services (World Bank, 1984).

Table 8.1 shows the major crops in Cameroon. The country is an important producer of several major tropical export crops, and is one of the few countries that produces considerable amounts of both arabica and robusta coffee. About 57 % of total exports consists of agricultural and agro-industrial products, of which coffee with 19 %, cocoa with 18 % and cotton with 4 % of total exports are the major commodities (FAO, 1980-1983). With an average annual production of 102 000 t in the period 1979/81, Cameroon contributed 2 % of the world production of coffee, and ranked 12th as world producer just after Costa Rica and before Peru, Kenya and Madagascar.

8.2 Ecology

Stretching for more than 1200 km from Lake Chad to the Gulf of Guinea, Cameroon has roughly the shape of a triangle, with the baseline on 2°N and pointing north to 12°N. The 'Cameroon Dorsal', a region of highlands crossing the country from the south-west in northeastern direction, separates the South Cameroon Plateau with an average altitude of 700 m, from the northern lowlands which are on

Table 8.1. Average area harvested, yield and national production of major crops over 3-year periods.

Crops	Area (1000 ha)		Yield (kg/ha)		Production (1000 t)	
	1969/1971	1979/1981	1969/1971	1979/1981	1969/1971	1979/1981
Food crops						
cassava	161	233	3965	4308	637	1004
maize	309	542	1151	860	355	466
plantain					694	988
sorghum & millet	477	466	719	871	343	405
yam & taro	259	363	3328	3366	862	1222
Oil crops						
groundnuts	244	353	845	320	206	113
oil palm	.	55	.	2273	104	125
Beverages						
coffee	241	352	373	292	90	102
cocoa	387	426	328	276	127	118
Other crops						
cotton	103	57	561	1374	58	78
rubber	21	29	620	590	13	17
sugar-cane	11	20	909	2300	10	46

Source: FAO (1971-1983)

average only 300 m above sealevel (Figure 8.1). Average temparatures are generally high, with little variation during the year, but these vary from about 20 °C on the 'Dorsal' at 1400 m altitude, through 24 °C on the southern plateau, and 26 °C along the coast, to about 28 °C in the north. The amount of sunshine varies in a similar manner from about 1000 hours in the south-west to about 3000 hours per year in the north.

The climate is affected by two air masses: in the south by the unstable and moist monsoon or southeastern trade winds, and in the north by the stable and dry harmattan or northeastern trade winds. During the winter period (December to January) the harmattan invades the north and the central part of the country, resulting in a dry season, whereas in summer the monsoon moves up to the north and brings the rainy season with it.

Because of the relief and the proximity of the ocean in the south-west, the average annual rainfall varies considerably, from more than 4000 mm along the coast, about 3000 mm in the west along the Nigerian border, 1500-2000 mm around the Dorsal and in the south-east to less than 1000 mm in the north, and even less than 500 mm around Lake Chad (Table 8.2 and Figure 8.2).

On the basis of these climatic variations the country is divided into three main climatic regions. The first, the equatorial region covers the whole of the south, and is subdivided into: the Guinean type (four different seasons) and the Camer-

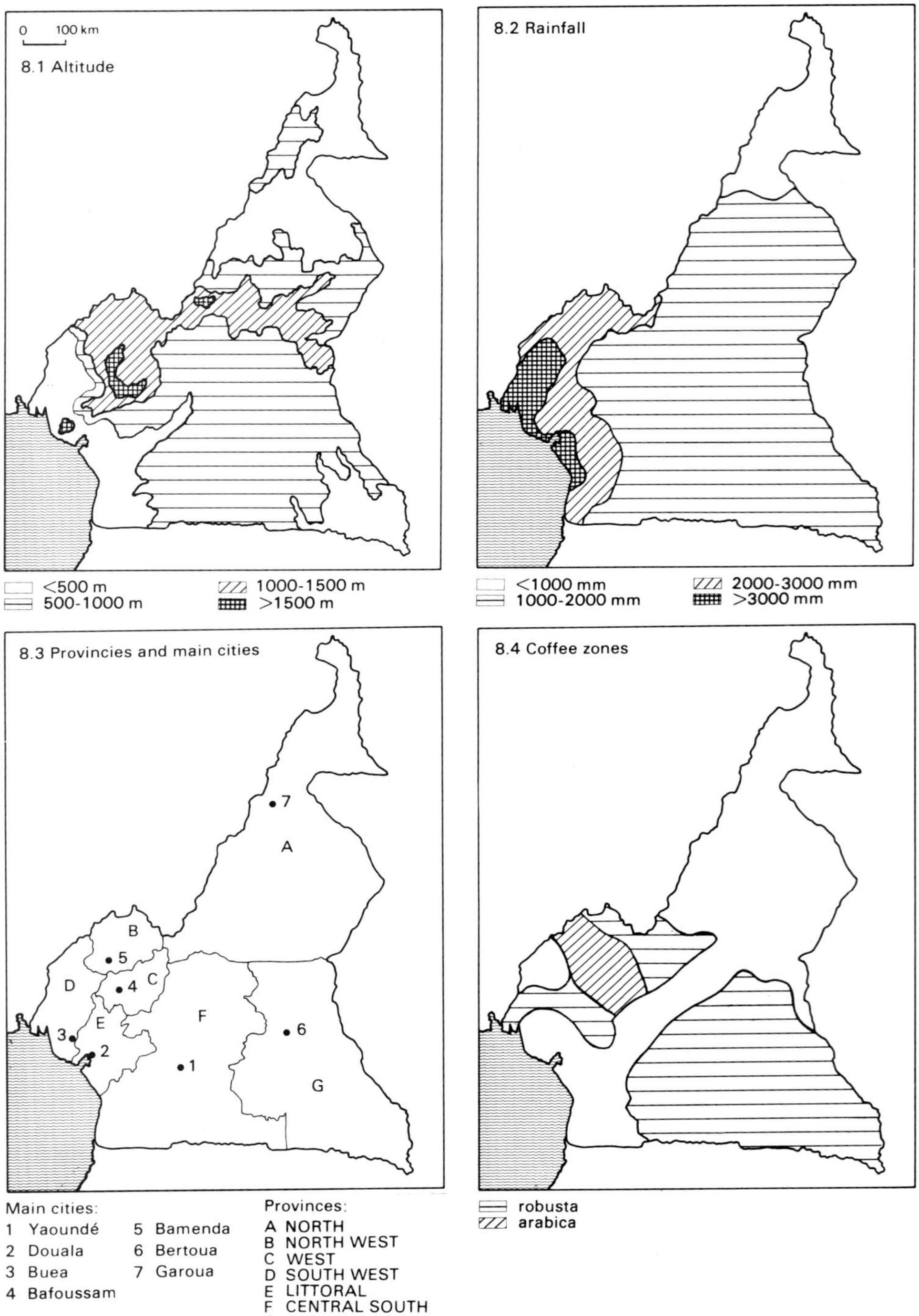

Fig. 8.1-8.4. Altitude (8.1), average annual rainfall (8.2), provinces and major cities (8.3) and suitable areas for coffee (8.4) in Cameroon. After Les atlas jeune afrique (1980).

Table 8.2. Distribution of rainfall of zones with coffee.

Province	Station	Altitude (m)	Annual rainfall (mm)	Monthly rainfall (mm) J	F	M	A	M	J
NW	Bamenda	1524	2635	29	53	153	205	228	323
W	Dshang	1398	1919	21	49	137	189	188	236
	Foumbot	1100	1719	8	27	96	142	164	172
SW	Manfé	152	3410	32	69	175	220	321	428
Lit	Douala	13	4125	58	84	223	243	337	498
CS	Yaoundé	759	1587	27	62	144	184	211	144
E	Bertoua	670	1564	24	53	111	142	158	175

Station	J	A	S	O	N	D
Bamenda	412	370	489	270	79	24
Dshang	228	241	333	237	48	12
Foumbot	224	252	305	250	67	12
Manfé	497	454	559	466	149	40
Douala	707	767	625	383	148	53
Yaoundé	57	77	222	302	134	23
Bertoua	98	135	232	297	108	32

Source: Ministère de l'Agriculture (1981)

oonian type (with abundant rainfall throughout the year). The Sudanian climate, with 3 to 6 dry months, is found in the belt from 7° to 10°N, and the Sudano-Sahelian climate with 7 or more dry months is found in the most northern part of the country.

The West and North-west Provinces offer on the highlands excellent climatic conditions for arabica coffee, as long as the annual rainfall remains above 1500 mm and the length of the rainy season exceeds 9 months. Robusta coffee can be grown in a large area, including the South Cameroon Plateau (Central South and East Provinces) and in some parts of South West and Littoral Provinces in the Cameroonian equatorial climatic region (Figure 8.4).

Cameroon not only offers a wide climatic variation, but it also has many different soil types. In the climatic zones suitable for robusta, ferralsols are generally found, which are usually clayey, porous, well developed and often covered with forests. Although poor in nutrients and with little humus, these soils are suitable for coffee. The Western Highlands, which constitute the arabica zone, contain both red mineral soils as well as aliatic soils with a fine grained structure and a hardened horizon, which by erosion might come to the surface. These soils are also suitable for coffee, but the best soils are the lightly evolved soils of volcanic origin (andosols) and the eutrophic brown soils, which appear in some places in and around West Province (Bouharmont, 1982). Those combine a good structure with a high natural fertility.

8.3 The coffee sector

8.3.1 History

Coffee was introduced in Cameroon in 1905, then a German colony under Governor Von Puttkamer. Around that time several major infrastructural works were constructed, such as the Kribi Yaoundé road, the northern railway line from Douala up to Nkongsamba and the first port facilities at Douala. Development of the infrastructure continued under the French mandate and in 1927 the Central railway line reached Yaoundé. These infrastructural works favoured cultivation of export crops, and from 1930 onwards cocoa and to a lesser extent coffee became increasingly popular.

The plantations were established in the densely populated areas of West and North-west Provinces and in Moungo Department of Littoral Province, along the Douala-Bafoussam road. During the last few centuries these areas had attracted a large population, due to the naturally favourable environment (high altitude climate, partly volcanic soils, etc.). These conditions are also suitable for coffee, particularly for arabica, which furthermore requires considerable labour inputs, and therefore fits well in densely populated areas. During the last decennia robusta coffee has expanded into several other areas such as the Central South and East Provinces.

8.3.2 Recent developments

Although the respective sources give different statical data in particular of the area cultivated, the increase of the coffee production over the period 1960 to 1980 is to be attributed solely to the rapid growth of the area under coffee, from just over 120 000 ha in 1960/61 to about 370 000 ha in 1980/81. The total production increased during that period from 51 000 t to 112 000 t (Table 8.3).

Development of the arabica production was somewhat different from that of robusta. During the 1960s the area under arabica increased considerably from 35 000 ha to about 110 000 ha, and the production increased accordingly. Since 1970 the area increased at a much slower pace and average yields dropped to only 170 kg green coffee per hectare. Development of robusta on the other hand was more pronounced in the 1970s, when the area cultivated increased from about 120 000 ha to 210 000 ha and the production rose from 50 000 to 85 000 t. Average yields remained around 400 kg green coffee per hectare.

Cameroon has several distinct production zones of coffee. The production of arabica is concentrated in the North-west and West Provinces, at higher altitude. And the major areas of robusta production are the northern part of Moungo Department in Littoral Province and areas adjacent to that, in West Province (southern part) and in South-west Province (Table 8.4 and Figure 8.4). Given the agro-ecological conditions, the simple processing methods and the modest labour re-

Table 8.3. National production of coffee and cocoa (1000 t) (1960-1982). Figures in brackets refer to area in 1000 ha.

Crop year	Arabica	Robusta	Total coffee	Cocoa
1960/61	10.2 (35)	41.0 (85)	51.2 (120)	77.0
1970/71	26.2 (110)	50.0 (120)	76.2 (230)	113.2
1971/72	30.5	63.4	93.9	124.2
1972/73	31.3	62.9	94.2	106.9
1973/74	24.7	67.0	91.7	110.5
1974/75	31.1	80.0	111.1	119.3
1975/76	22.0	57.2	79.2	104.6
1976/77	18.9	62.1	81.0	81.7
1977/78	21.5	64.4	85.9	101.9
1978/79	27.0	80.0	107.0	105.8
1979/80	31.5	83.3	114.8	121.9
1980/81	27.1 (160)	85.3 (210)	112.4 (370)	119.5
1981/82	25.4	86.7	112.1	120.2
1982/83	22.2	104.3	126.5	106.1

Sources: CNCC (1980a), Ediafric (1983), Gicam (1983)

Table 8.4. Distribution of the area under coffee (1000 ha) and the production (1000 t) by Province (1980/81).

	Robusta		Arabica		Total	
	area	production	area	production	area	production
Littoral	52	37	–	–	52	37
Central South	31	7	–	–	31	7
East	28	9	–	–	28	9
North West	6	1	60	8	66	9
West	41	23	104	17	145	40
South West	50	10	–	–	50	10
Total	208	87	164	25	372	112

Source: Ministère de l'Agriculture (1981)

quirements there are still great possibilities for the expansion of robusta, in particular in Central South and East Provinces. In those areas it faces competition from cocoa.

8.3.3 *Organization*

Through the dispersion of the respective production zones, in which an estimated 330 000 farmers produce some coffee, the organization of production, processing and marketing is largely decentralized. The Government is not directly involved in the production of coffee, nor in the processing and marketing, but has delegated certain responsibilities for the coffee sector to the ONCPB (Office Nationale de Commercialization des Produits de Base), which is the legal owner of most marketed coffee.

This parastatal marketing board was established in 1976, when it took over the activities of the former 'Caisse de Stabilization'. Its main objective is the stabilization of primary export commodity prices in Cameroon (coffee, cocoa and certain oil crops). ONCPB plays a major role in the export of coffee and is, as a regulatory body, responsible for legislative and administrative policies related to marketing and pricing. It issues before every crop year the so-called 'Textes de Campagne', in which all marketing regulations are published, including such items as the licensed exporters of robusta coffee, the licensed buying agents and the official buying prices.

In 1983 ONCPB gave licenses to 23 exporters for the internal marketing and export of robusta coffee. ONCPB itself takes care of the marketing and export of

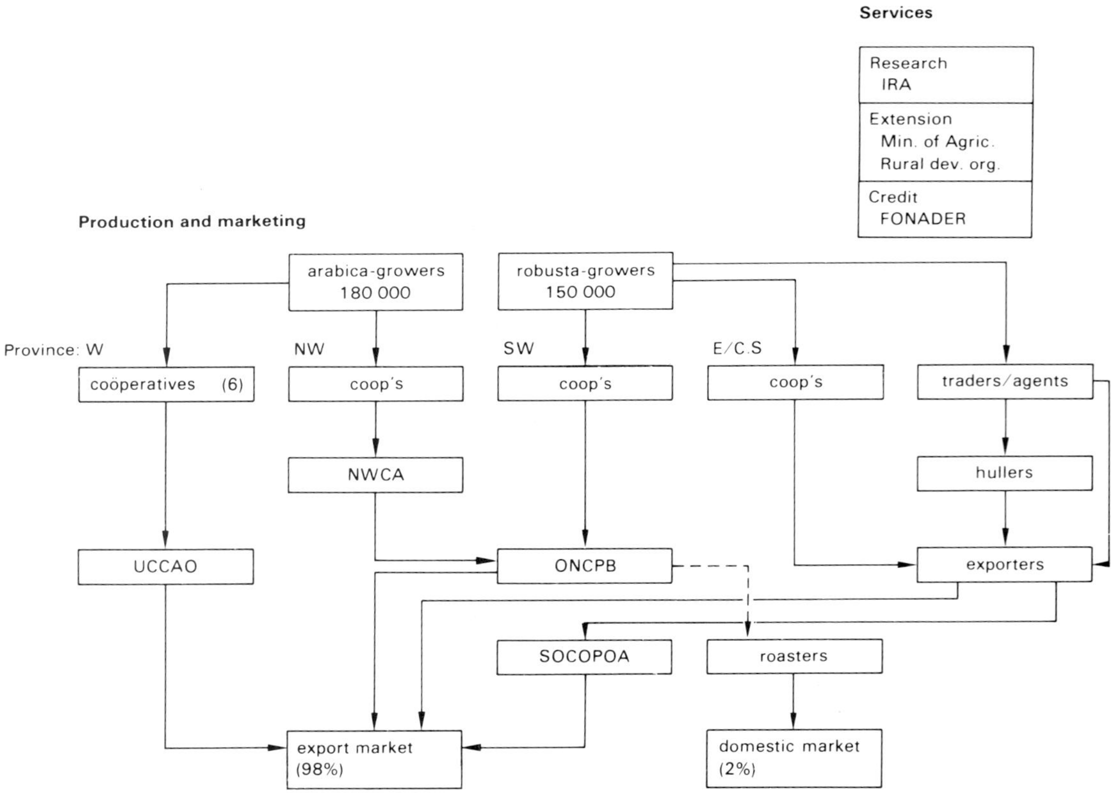

Fig. 8.5. Participants in the coffee sector and main marketing channels in Cameroon in 1982. Solid line, main channel; broken line, secondary channel.

the coffee from the North-west and South-west Provinces. In North-west Province it has appointed the NWCA (North West Cooperative Association Ltd.) as sole buying agent (arabica coffee). The arabica coffee of West Province is directly exported by the UCCAO (Union Centrale des Coopératives Agricoles de l'Ouest).

The producer prices which ONCPB establishes every year for the Government, are generally more in line with the cost of living index than with the prices on the world market. The reserves thus created in years of high prices on the world market are used to subsidize organizations involved in input supply and in the improvement of the rural infrastructure, and organizations responsible for programmes aimed at the diversification of agricultural production. Apart from the factories belonging to the large cooperatives, there are about 180 major hulling factories in the country, almost half of which are located in Moungo Department. Through the Institute of Agricultural Research (IRA) the Government is involved in coffee research activities, and the Ministry of Agriculture is largely responsible for extension services to coffee farmers.

8.4 Production

8.4.1 Production units and farming systems

According to the latest agricultural census of 1972/73, about 925 000 families were then involved in small scale agriculture, which consists largely of mixed plantings of food crops and perennials, such as coffee, cocoa and oil palm. The average farm size was about 1.6 ha, with the smallest farms in the densely populated areas of West and North-west Provinces and slightly larger farms in Central South and East Provinces (Table 8.5). There is, however, a wide range of farm sizes, and given the complicated land tenure systems with many distinct land rights, which differ in particular between men and women, it is difficult to define and measure the size of a farm. In the past ten years the number of households has increased considerably, whereas the cultivated area expanded relatively more in the less densely populated areas.

Farming systems are quite different for the respective production areas and are related to the staple food crops which vary according to the ecological zones. In the north sorghum is the major food crop. In south ward direction it is increasingly found in combination with maize and cassava. In the Central South and East Provinces plantains are important, together with such crops as yam and groundnuts. Taro and cassava are the most frequently found food crops in the South West Province, whereas the West and North West Provinces are the main suppliers of maize and vegetables.

A typical farm with robusta in the forest/savanna zone would show some fields with cassava and plantains, a field with groundnuts followed in a rotation by maize or pumpkins, and some fallow fields, all of which managed by the women. Closer to the homestead, usually along the road, are the cocoa trees and the coffee bush-

Table 8.5. Agricultural production units and area cultivated by province (1972/73).

Province	Number of farm households (× 1000)	Agric. work force (× 1000)	Cultivated area (× 1000)	Av. farm size (ha)	Area under coffee* (1000 ha)	Proportion coffee area of cultivated area (%)
North	312	808	511	1.64	–	–
Nort West	110	315	134	1.22	50	37
West	129	359	161	1.25	120	74
South West	79	201	115	1.46	35	30
Littoral	50	126	75	1.49	35	47
Central South	189	514	381	2.02	20	5
East	56	155	102	1.81	20	20
Total	925	2478	1479	1.60	280	19

Source: Agricultural Census (1972/73), quoted by World Bank (1980)

es, which are looked after by the men. In the Bamiléké country in the Western highlands, a major arabica zone, the pattern is different: the houses are scattered in their small individual concessions along the slopes. These are covered by many different plots some of which only planted with food crops (maize, beans and vegetables) and others with the same food crops grown under arabica coffee and banana trees. The women cultivate all food crops, whether on separate plots or mixed with coffee. With the demographic pressure and the extension of the coffee plantations, grazing lands have become scarce. Because of the high and further increasing population density in the arabica zone, food crops have first priority and the coffee in between those food crops is neglected, whereas coffee plantations in the robusta areas are on separate plots and have a place of their own within the farming systems.

There are still about 15 large coffee estates, formerly managed by Europeans, and grouped together under COOPAGRO.

8.4.2 *Production stages and production costs*

Since such agronomic parameters as variety, planting density, cultivation practices and average yields, do not vary much between the respective production zones, information on production costs is only given for the major production zone: Western Province. This province produces two-thirds of the arabica coffee in the country and more than a quarter of the national production of robusta. It is one of the few places in the world where production zones of arabica and robusta are adjacent to each other, which makes it possible to compare their costs of production, under more or less similar circumstances. The ecological conditions are somewhat different, with arabica coffee generally above and robusta coffee below

Table 8.6. Establishment costs (year 1-4) per hectare coffee (arabica or robusta) in the Western Province (1982/83).

Cost elements	1	2	3	4	Total
Labour inputs (man-days)					
clearing & lining out	35	–	–	–	35
digging of planting holes	80	–	–	–	80
planting	15	–	–	–	15
weeding & cleaning	15	25	25	25	90
fertilizing	10	12	13	15	50
pruning	–	–	–	7	7
total	155	37	38	47	277
Subtotal (CFAF 1000)					193.9
Material & other costs (CFAF 1000)					
plants (CFAF per plant)	6.2	0.3	–	–	6.5
fertilizers	1.8	3.5	5.3	7.0	17.6
transport plants	5.0	0.2	–	–	5.2
transport manure	–	3.0	4.0	5.0	12.0
Subtotal (CFAF 1000)	13.0	7.0	9.3	12.0	41.3
Total establishment costs (CFAF 1000)					235.2

Source: UCCAO (1983)

1000 m altitude, and the processing techniques are as usual: wet processing for arabica, and dry processing for robusta coffee. The comparison is to some extent blurred by the distinct farming systems adopted for these two types of coffee: arabica coffee is generally intercropped with food crops, and robusta coffee is only partially associated with cocoa and some other tree crops.

The establishment period (4 years) and costs are, however, similar for both types and are here considered the same (Table 8.6). The normal productive life is somewhat less for arabica than for robusta (20 years against 25 years). Table 8.7 shows the annual costs of production of arabica and robusta coffee, under traditional and improved cultivation practices, respectively. The production stages and activities are described hereunder.

The planting density for arabica coffee varies between 1000 to 1500 trees per hectare, depending on soil type and intensity of intercropping with food crops. The age of coffee trees is generally high, with about 50 % above 20 years. One-stem pruning is applied, and weeding is more specifically undertaken for food crop production. Some crop residues and manure are used for fertilization, but chemical fertilizers are also applied at a rate of about 200 kg/ha (meant for food crops). Pest and disease-control concerns the spraying against coffee berry disease (now undertaken by 30 % of the farmers) and antestia (*Antestiopsis lineaticollis intricata*) undertaken by the Pest Control Unit of the Ministry of Agricul-

Table 8.7. Production costs of (green) arabica and robusta coffee (per hectare and per kilogram) in the West Province (1982/83).

	Arabica		Robusta	
	traditional	improved	traditional	improved
Yield (kg/ha)	250	325	425	625
Plant density (plants/ha)	1250	1250	1300	1300
Productive period (years)	20	20	25	25
Wage rate (CFAF/man-day)	700	700	700	700
Annuity of establishment costs (CFAF 1000) (10 %)	27.5	27.5	25.8	25.8
Annual costs				
Labour inputs (man-days)				
weeding	20	20	40	50
fertilizing	7	10	4	10
disease control	8	15	4	8
pruning	7	9	10	15
harvesting	30	34	33	44
processing	6	7	–	–
drying	3	3	4	5
transport	3	3	5	6
total	85	101	100	138
Subtotal of costs (CFAF 1000)	59.5	70.7	70.0	96.6
Material and other costs (CFAF 1000)				
fertilizers	7.0	14.0	3.5	14.0
chemicals	4.0	10.0	1.8	3.6
processing materials	6.0	8.0	2.0	3.0
depreciation tools	7.0	10.0	7.0	10.0
transport manure, etc.	2.0	6.0	–	–
other	–	–	6.0	9.0
Subtotal	26.0	48.0	20.3	39.6
Total costs per hectare (CFAF 1000)	113.0	146.2	116.1	162.0
Costs per kilogram (CFAF)	452	450	273	259
(US$)	1.26	1.25	0.76	0.72

Source: UCCAO (1983)

ture. The harvest takes place in the period October to December and is undertaken by the men, with assistance of children.

Yields are generally low (not more than 250 kg green coffee per hectare) which can be attributed to the intensive intercropping of food crops and the age of the trees. After harvesting the farmer transports his cherry coffee to the local water point, where it is pulped, fermented and washed. He then takes it back to his house where it is dried on the ground or on cement platforms. A small amount of arabica coffee is processed according the dry method.

In the end the parchment coffee is brought in 60-kg jute bags to the cooperative purchase centre for sale, from where it is eventually transported to the hulling plant.

The planting density for robusta is approximately 1200 to 1500 trees per hectare and depends on the intercropping with cocoa and other trees. The age of coffee trees is usually high, and the extent of new plantings is insufficient to reduce the average age of plantations. Weeding is undertaken irregularly, usually by the men. Many robusta plantations do not receive any fertilizer at all, and the average amount of fertilizers applied does not exceed 100 kg/ha (50 % sulphate of ammonia; 50 % compound fertilizer: 20-10-10). Pest and disease-control is not important and is mainly aimed at the protection against the coffee berry borers, other boring insects and caterpillars.

The harvest takes place in the period December to February, and is followed by sun-drying of the whole cherries, which often takes as much as 8 to 10 weeks. The farmer then delivers his dried cherries to the cooperative to be hulled.

The yields of robusta are considerably higher than those of arabica, mainly because of the absence of intercropping with food crops and although robusta in Western Province is often found under marginal ecological conditions (relatively low temperatures and low rainfall). The yields vary between 400 to 450 kg green coffee per hectare, and could with better cultivation practices reach 600 to 650 kg/ha.

8.4.3 Yields and returns

The previous section shows that the present average yields of coffee are low, in particular for arabica coffee. The underlying factors are also discussed. Whereas the average production costs per hectare for traditional plantations are for both arabica and robusta coffee around CFAF 115 000 per hectare (1982/83), the average production costs per kilogram green coffee are not less than CFAF 450 for arabica coffee against only CFAF 270 for robusta. The official producer prices (per kilogram green coffee) in that year were CFAF 370 for arabica and CFAF 350 for robusta (for the year 1983/84 these were raised to CFAF 410 and CFAF 390). Given the production costs, and although arabica growers receive an additional payment at the end of the season ('ristourne'; about CFAF 30 to 50 per kilogram), the growing of robusta is certainly more attractive.

In Table 8.7 labour inputs are valued against the average wage rate in these rural areas, which was CFAF 700 per man-day in 1982/83. But in fact most of these labour inputs concern family labour, and the relatively high production costs not necessarily result in financial losses. However the net return to (family-)labour is under traditional cultivation practices, only about CFAF 460 per man-day for arabica coffee against not less than CFAF 1000 per man-day for robusta coffee.

Because of the population pressure and the social need and economic benefits of food crop production, it is unlikely that intercropping of arabica coffee will be abandonned. Hence there is little prospect for a considerable increase in arabica yields, in West Province. But through the introduction of the new Java variety, and other measures, some expansion of arabica can be expected in North West Province. Average yields of robusta coffee can be raised easily, not only in West Province but also in surrounding areas and other suitable zones for robusta coffee, and those higher yields would further reduce the average production costs per kilogram, as shown in Table 8.7.

8.5 Processing

The processing of both arabica and robusta coffee is undertaken by a multitude of individuals and organizations, using different methods and equipment. Hereunder the main processing stages and methods are discussed.

A major problem in the farm level processing of arabica, which also affects other processing stages is the inadequacy of pulping facilities, resulting in a delay of more than two days between harvesting and pulping. There is also an insufficient supply of clean water at these centres. (Per tonne coffee cherries about 10 m^3 is required.) The use of water that is coloured by the red soil, results in red parchment which makes it difficult at a later stage to eliminate stinking beans, through both hand and electronic sorting.

Arabica coffee is generally handled by cooperatives, which store it, often in tanks due to a lack of bags, and then bring it to the cooperative processing plant. The processing includes the following operations: pre-cleaning, to remove foreign matter; hulling; polishing; grading; mechnical cleaning, to remove dust and defective beans, and finally sorting. The machinery is generally not well maintained and for various reasons not very effective. Both the pre-cleaning and grading machines have screen frames, made of wood, which tend to swell due to the high humidity, and cannot be removed for the necessary cleaning.

The maximum capacity of most cooperative factories is about 36 t per 24 hours (4000 t per year) in West Province and 6 to 24 t per 24 hours in North West Provinces. In the last province the equipment is quite old. A major bottleneck constitutes the sorting which is still mainly done by hand, at an average productivity of 60 to 120 kg per man-day. For this activity cooperatives employ between 50 to 500 seasonal labourers, which received in 1980 not much more than CFAF 250 per 60-kg bag. The average costs of hand sorting under these circumstances is nevertheless in the order of CFAF 5 to 8 per kilogram against a total cost price of electronic sorting, including cleaning and bagging of CFAF 16 per kilogram. The Central UCCAO processing factory has 20 biochromatic electronic colour sorters, each with a capacity of more than 10 t per 24 hours, and requiring an investment outlay of US$25 000. This equipment is irregularly used e.g. when primary cooperatives do have a shortage of labour or when extremely dirty coffee has to be sorted.

There are six grades (A to F) including peaberries (C) and elephant beans (E), and excluding damaged and broken beans. The larger beans (grades A to C), of which relatively few are produced, usually fall into the highest quality class (Prima; less than 30 defects per 300 gram) followed by D, as Superior class (60 defects) and F as Current class (90 defects). The caffein content is on average 1.27 %.

Robusta coffee is processed by the dry method, with the advantage that after harvesting, besides drying no immediate other steps have to be taken. Drying

Table 8.8. Costs for robusta processing plants with a capacity of 12000 t per year (1982).

Costs elements	Integrated processing (all processingstages in one factory)		Decentralized processing (5 hulling factories & 1 grading/sorting plant)	
	number	US$1000	number	US$1000
Hulling stage				
bridge weigher	1	21	1	10
dryer	1	198	1	70
precleaner	1	38	1	20
stoner	2	21	1	12
cherry separator	1	41	1	15
hullers	1	20	1	18
elevator & pneumatic transport	5/4	52	1/1	12
electronical installations	.	.	.	85
other equipment	.	.	.	28
installation, freight insurance	.	.	.	130
total for each hulling factory				400
Grading and sorting stage				
bridge weigher	.	.	1	30
grader	1	66	2	46
gravitycleaner	1	94	1	64
elevator & pneumatic transport	1/4	26		
colour sorters	8	200	12	360
silos & tanks	.	41	.	.
weigher and sewing machine	2	19	2	14
electrical installation	.	132	.	100
other equipment	.	181		236
installation, freight, insurance	.	450	.	350
Total for separate grading/sorting plant				1200
Total integrated processing		1600		
Total decentralized processing				3200

Sources: FAO/World Bank (1981), Data from project ZAPI-EST

takes often more than 6 weeks and even then the moisture content might still be about 14-16 %, whereas for hulling 12 % or less is required. From the start of the buying season, in early January, bags are distributed to the farmers by the cooperatives, agents and processing agencies, for the collection of the dried cherries. The processing stages are similar to that of arabica, but the cherries have first to be dried completely and to pass a cherry separator. There are more than 200, mainly private, hulling factories processing robusta, with an average turnover of only 400 t green coffee per year (one season per year, from January to April). Many of these small plants are situated in one department (Moungo), where the cooperative movement was not successful in competing the private exporters and their agents.

Recent studies undertaken for the establishment of processing facilities in West and East Provinces show that large installations with a total capacity of 5 t/hour or 12 000 t/year require an investment of at least US$2 to 3 million (Table 8.8). Costs of depreciation and interest alone therefore amount to about CFAF 8 per kilogram. The personnel required for one integrated factory is about 20 staff members and 150 to 200 unskilled labourers. This represents again a cost of about CFAF 8 per kilogram (2 to 3 dollar cents). Other important processing cost elements are maintenance, electricity and bags and other materials. The problems of hulling, grading and sorting of robusta are similar to those for arabica coffee. There are a few large industrialized grading and sorting factories. One of these, in Nkongsamba, belongs to SOCOPAO, and that factory, until recently, also treated the robusta coffee for UCCAO (in 1980 at a fee of CFAF 13.20 per kilogram).

For robusta four grades are distinguished, large beans (screens 18 and above) and grade I, II and III, respectively, and the classification is similar to that of arabica, with one additional class: Extra Prima (up to 15 defects per 300 gram).

About 13 firms are involved in roasting and grinding of coffee for domestic consumption. There are no instant coffee factories in Cameroon.

8.6 Marketing

8.6.1 Marketing channels and functions

The marketing channels of coffee vary considerably not only according to the type of coffee concerned (arabica or robusta) but they also vary by region and by the extent that cooperatives are involved. For arabica two marketing systems can be distinguished: the production of West Province is handled and exported by the Cooperative Union (UCCAO) itself, whereas the arabica coffee from North-West Province is handled and processed by a similar organization, the NWCA, on behalf of the board (ONCPB). The Board takes care of the export of all the coffee from the North-West as well as the South-West Provinces (Anglophone Cameroon). For the marketing of robusta in the francophone areas, the Government (ONCPB) every year establishes a list of exporters (now 23), and the proportion of expected production these exporters are entitled to purchase. They are penalized if they remain more than 15 % below or if they go 15 % or more beyond their quotum. On this list feature several cooperative organizations, such as UCCAO, Zapi-Est and Cooplam, several large processing companies and a certain number of large trading firms, which obtain the coffee through cooperatives and agents.

Apart from collection activities, processing operations and export arrangements, major marketing functions consist of storage and transport. Storage appears to be a major problem, for both unprocessed and processed coffee. The problem is aggravated by the fact that two types of coffee with a total of more than ten grades (including damaged beans) are involved. The coffee is usually stored in quantities of 250 bags, a measure by which it is also exported. The total storage capacity in the arabica zones (UCCAO and NWCA) was in 1980/81 about 150 000 m^2, or about 30 000 t. This is about one year production of coffee, but other produce has to be stored too. The major problem lies in the delay of sales operations and the lack of storage information systems. Similar problems exist for robusta coffee, but there are no accurate data on total storage capacity. In recent years the storage capacity had to be enhanced, because of the rapidly increasing stocks of unsold coffee.

The coffee that is ready for shipment is transported by UCCAO, Zapi-Est, COOPAGRO and other exporters to the major harbour, Douala, where it is stored by SOCOPOA; free of charge during the first twenty days, and at a rate of 32 CFAF per tonne per day thereafter. SOCOPAO, which belongs to a French multinational organization, performs several marketing functions for both coffee and cocoa. Its storage capacity in Douala is about 50 000 m^2 (1982) which, given an average storage period of more than one month and the space required for cocoa, is not enough for the coffee. Exporters often need to hand over their coffee to SOCOPAO, to obtain credit from the bank, but on the other hand they wish to reduce the period the coffee is stored by SOCOPAO, because of the costs involved and the humid climate in Douala. Rates for long term storage in Douala (Office National des Ports) were in 1980 CFAF 5000/m^2 per year in a warehouse and CFAF 1350/m^2 per year in the open air (to be multiplied by 5 to obtain costs per tonne per year).

For the transport of coffee, exporters require a 'lèttre de voiture', to be signed by the processing agency, the local government (Autorité Administrative) and the military police (Gendarmerie). Arabica coffee is only transported by road, whereas a large amount of robusta coffee is transported by rail, although the amounts transported by rail from Nkongsamba to Douala diminished in recent years. The costs of transporting coffee by rail from Belabo (East Province) to Douala, over a distance of about 400 kilometres, amounted in 1980 to about CFAF 11 000 per tonne. That coffee is first transported by road, to Belabo, over an average distance of another 100 kilometres. That was in 1980 undertaken by ZAPI-EST, at a cost of about CFAF 4500 per tonne. The ONCPB, Department of Purchases, every year establishes a list of allowed transport rates per tonne of robusta, for the respective production areas. These rates range from about CFAF 5000 per tonne transported from Moungo-department to Douala, to not less than CFAF 50 000 per tonne from the Mouloundou area in East Province.

From the above it follows that both storage and transport are costly and do require much coordination which often goes beyond the capabilities of the respective organizations involved.

8.6.2 The role of coffee cooperatives

Cooperatives play a major role in the marketing of coffee and cocoa in Cameroon. In some areas, e.g. West Cameroon, cooperatives already have a long history, whereas in other areas, they have only recently been introduced. There are a total of about 50 coffee cooperatives, 12 of which can be considered large cooperatives, with a membership of several thousand and a volume of produce handled of several tonnes. As indicated already in the previous sections the cooperatives in the respective production areas perform different marketing functions. The six large cooperatives in West Province and their union, the UCCAO, are involved in all marketing operations, including export, whereas the cooperatives in Central South Province are only involved in the collection of robusta coffee, on behalf of exporters.

In 1982 a detailed study was carried out on the financial results of coffee and cocoa cooperatives in Cameroon (SEDES, 1982). In the following some information is presented about the revenues and costs, incurred for the major activities by a sample of the cooperatives, covered in that study. Four cooperatives are included from West Province and one cooperative from each of the other coffee-producing provinces. The townships in which these cooperatives are situated, the volumes handled, and the revenues and costs for the year 1980/81, are presented in Table 8.9. The largest UCCAO-cooperatives: Caplamé (Dshang) and Caplami (Bafoussam), which are mainly involved in the marketing of arabica coffee, did obtain in 1980/81 high net results, whereas another UCCAO cooperative: Caplahn (Bafang) was unable to cover its costs with the commission obtained for the handling and processing of robusta coffee. It is interesting to note that the costs per kilogram coffee handled are relatively close for those (7) cooperatives that are involved in both marketing and processing of coffee. (The two last cooperatives, in Central South and East Provinces, are only involved in the collection of coffee and their costs are low.) But the revenues per bag of coffee, obtained by the respective cooperatives vary enormously. The margins allowed for the marketing of robusta, as presented every year in the so-called 'Barème-ONCPB' were apparently insufficient for cooperatives handling that type of coffee. The above mentioned study therefore made suggestions to revise the official margins to remunerate cooperatives and exporters more evenly, more in line with the costs they incur.

Table 8.9. Revenues and costs of selected coffee cooperatives (province between brackets) (1980/81).

	Cooperatives[1] in								
	Bam (NW)	Bfo (W)	Dsch (W)	Bfa (W)	Ban (W)	Nko (Lit)	Muy (SW)	Ako (CS)	Aboa (E)
Membership (× 1000)	.	22	26	4	4	.	.	.	.
Volumes handled (1980/81) (1000 t)									
arabica	0.6	3.7	5.0	0.0	0.1	–	–	–	–
robusta	–	0.1	1.5	3.0	1.4	2.0	0.8	2.1	2.0
cocoa	–	–	0.0	0.1	0.3	–	1.1	3.4	1.9
Total revenues (CFAF million)	57	871	791	54	93	84	95	43	27
Costs (CFAF million)									
wages personnel	20	105	120	53	37	22	16	16	10
transport	8	87	115	22	33	28	14	6	13
packing materials	5	28	113	4	10	–	4	–	–
conseil d'administration	2	12	21	–	5	2	3	–	2
services, nurseries & training	5	38	23	58	10	3	10	15	4
costs of hulling	2	16	32	–	7	8	3	–	–
shares, subsidies & interest	1	–48	–46	–	–1	–	–2	–	–7
profits/losses	–	–42	34	14	–26	2	9	–9	–1
depreciation hulling installations	–	7	13	2	5	5	3	–	–
depreciation other capital	2	8	20	2	4	–	2	3	1
total	45	211	445	155	84	69	62	31	22
Net results	12	660	346	101	9	14	33	12	5
Bonuses to farmers	12	83	88	3	5	7	16	17	10
Revenues (CFAF per kg)	91	232	122	17	52	42	50	8	7
Costs (CFAF per kg)	72	56	68	48	47	35	33	6	6

1. Abbreviations that stand respectively for: Bamenda, Bafoussam, Dshang, Bafang, Bangangte, N'kongsamba, Muyuka, Akonolinga and Abong M'bang.
Source: SEDES (1982)

The importance of the UCCAO-complex (Headquarters and 6 cooperatives) can be illustrated by its total membership of about 97 000 farmers and the total number of employees of 4300 of which 1200 are permanent. Each of the cooperatives has several sections, composed of a certain number of centres. E.g. the cooperative Caplami has 7 sectors and a total of 55 centres. The UCCAO cooperatives altogether own at least forty lorries, ranging from 7 to 20 t.

8.6.3 *Marketing costs and margins*

In the previous sections the respective cost elements in the processing and marketing of coffee have been discussed. Table 8.10 gives the respective marketing margins, as these are established before every season by the ONCPB, in the so-

Table 8.10. Prices, marketing costs and margins in CFAF and US$ per tonne of robusta and arabica (green coffee equivalents) (1982).

	Robusta		Arabica	
	(CFAF 1000)	(US$ 1000)	(CFAF 1000)	(US$ 1000)
c.i.f.-price (France)	803	2.44	921	2.80
Costs of shipment				
freight, handling	30		30	
insurance, finance (2.66 %)	21		25	
quality discount, etc.	44		52	
Total	95		107	
F.o.b.-price (Douala)	708	2.15	814	2.47
Costs of export				
customs, harbour charges	13		13	
export tax (32 % of v.m.)	66		77	
various other taxes	6		10	
margin exporters	4		5	
Total	89		105	
Gross earnings	619	1.88	709	2.16
Marketing costs				
collection, handling	} 31		26	
hulling, polishing				
grading	6		} 40	
sorting	8			
storage and losses	10		7	
transport	20		10	
cost of finance	27		20	
administration costs	10		35	
'prelèvement' ONCPB	60+		38+	
Total	172		176	
Net earnings	447	1.36	533	1.62
Price to producer	330	1.00	335[1]	1.11
Ristourne to producer	–		30	
Margin to reserves	117	0.36	168	0.51
Share of unit value of exports (f.o.b.) obtained by producers	47 %		45 %	

1. Allowance is made for low grade coffee, for which producers receive less than the official price of CFAF 350 per kilogram green coffee.
Source: SEDES (1982)

called 'Différentiel Café'. Export taxes are based on the 'Valeur Mercuriale', which in 1981/82 was CFAF 205 for robusta and CFAF 240 for arabica, and the rate applied was 32 %. There are also several other minor taxes, generally to be paid by the exporters. The fee to exporters for robusta and arabica coffee were CFAF 3500 and CFAF 5000 per tonne, respectively.

On the basis of the information on marketing margins, provided in Table 8.10, it is posible to estimate how the gross export earnings from coffee were distributed in 1982 among the respective participants in the production and trade (Section 8.8). Table 8.10 also shows that a considerable margin remains unexplained: that part of the export earnings from coffee that is partly reserved for price stabilization policies, and partly used to finance a large number of development projects and financial and industrial institutions in the rural and modern manufacturing sectors. The yearly earnings obtained by the Government from the export of coffee, including the taxes, the 'prelevement ONCPB' and the reserves, amounted in recent years to almost CFAF 25 thousand million. Similar amounts are probably obtained from the cocoa sector. Out of these funds the ONCPB, was able to provide a considerable amount of subsidies, in 1979/80 reaching about CFAF 40 thousand million. These subsidies are paid to the respective development organizations (Socapalm, Hevecam, Sodeble, Sodecoton, Mideviv, etc.) that are responsible for the diversification programmes, and to financial and other institutions (Fonader, Cocoa Roads, etc.) that are engaged in the improvement of the rural infrastructure. For Fonader these subsidies constitute about 60 % of its annual budget.

The question remains whether coffee farmers benefit indirectly or in the long

Table 8.11. Weighted averages of producer prices of robusta and arabica coffee, export unit values, general consumer price indices and share of export unit value obtained by producers (1973-1982) (index: 1973 = 100).

Crop year	Producer price		Export unit value		Consumer price index	Share of export unit value to producers (%)
	(CFAF/kg)	index	(CFAF/kg)	index		
1972/73	142	100	291	100	100	49
1973/74	149	104	331	114	117	45
1974/75	150	106	291	100	133	52
1975/76	170	120	459	158	146	37
1976/77	214	151	895	308	168	24
1977/78	269	189	816	281	189	33
1978/79	300	211	718	247	201	42
1979/80	321	226	808	278	220	40
1980/81	334	235	645	222	244	52
1981/82	341	240	757	260	276	45
1982/83	359	253	.	.	.	.

Sources: ONCPB (1983/84), IMF (1983)

run from the national price stabilization programme, which in most years reduces substantially their share in the total earnings from coffee. In Table 8.11 a comparison is made between the development of the weighted average producer price, the weighted average export unit value of coffee, and the general consumer price index. Table 8.11 shows that in the period 1973 to 1982 the export unit value had little influence on the fixation of the producer price. Except for the first few years the latter followed closely the general consumer price index. During the coffee boom period farmers received only 24 % of the export unit value. Despite these low payments farmers continued to pick and deliver their coffee, but they did not invest much in their plantations, a great many of which do urgently require rejuvenation or replanting.

8.6.4 *Export and domestic consumption*

The domestic consumption of coffee in Cameroon is modest, only 0.3 kg per person per year, and not less than 98 % of the coffee produced in the country is to be exported. During the 1970s, exports of coffee increased steadily, but in recent years they remained stagnant (Table 8.12). This is to some extent due to the quota system under the International Coffee Agreement, which limits exports from Cameroon to ICO-member countries to about 90 000 t, but also to the relatively modest sales to non-member countries. Since the national production was consistenly high during the last four years, the stocks of unsold coffee increased rapidly and reached in 1982 about 100 000 t, almost equivalent to one-year output. This contributed largely to the already mentioned storage problems for export commodities in the country.

Table 8.13 shows the proportional distribution of coffee exports by major importing countries. In the 3-year period 1978/81 the six countries mentioned im-

Table 8.12. National production, domestic consumption and exports (in 1000 t) (1960-1981).

Crop year	Opening stock	Production	Consumption	Available for export	Export	Closing stock
1960/61	2.8	51.2	0.1	53.9	44.9	9.0
1965/66	15.7	80.6	0.2	96.1	68.0	28.1
1970/71	28.7	76.2	1.0	103.9	60.0	43.9
1975/76	59.0	79.2	1.6	136.6	110.5	26.1
1976/77	26.1	81.0	1.6	105.5	73.7	31.8
1977/78	31.8	85.9	1.7	116.0	80.1	35.9
1978/79	35.9	107.0	1.9	141.0	93.7	47.3
1979/80	47.3	114.8	2.0	160.1	96.0	64.1
1980/81	64.1	112.4	2.2	174.3	91.6	82.7
1981/82	82.7	112.1	2.5	192.3	95.8	96.5

Sources: ONCPB (1983/84), CNCC (1980a)

Table 8.13. Proportional distribution (%) of coffee exports to major importing countries (1971-1981).

	1971-1974	1978-1981
France	29	24
United States	24	13
Netherlands	14	23
West Germany	13	13
Italy	5	17
Japan	5	4

Sources: CNCC (1980a), ICO (1977-1983)

ported not less than 94 % of the total coffee exports of Cameroon. Although none of these countries has the major share, the export market is nevertheless not very diversified. The major shipping company is CAMSHIP, with about 35 % of the total volume of coffee exports. Douala is by far the major port of shipment for coffee with 95 % of total exports. The other ports are Victoria and Kribi.

8.7 Supporting services

8.7.1 Research and extension services

Research activities in Cameroon are coordinated by the DGRST (Délégation Générale a la Recherche Scientifique et Technique), under which umbrella operate among others the ISH (Institut des Scienes Humaines) and the IRA (Institut de Recherche Agricole). IRA is responsible for agronomic research, whereas ISH among others undertakes research in the field of agricultural economics and rural sociology. Coffee research is carried out by IRA in collaboration with IFCC (Institut Français du Café et du Cacao), in several research centres in different parts of the country (IRA, 1981). The main centre is that of Nkolbisson, near Yaoundé in Central South Province at an altitude of 700 m. Other research centres for robusta coffee are located in East Province (Abong-Mbang) and in South West Province (Barombi-Kang), the latter at an altitude of 170 m only. The main centre for arabica coffee is in West Province (Foumbot).

For robusta, emphasis has been laid in the past twenty years mainly on the vegetative selection of high yielding trees (Bouharmont et al., 1979). Because productivity is much more important for growers than bean characteristics and organoleptic aspects, which are hardly reflected in producer prices, use has been made of 25 'clones prometteurs' since 1973 and of 8 'clones selectionnés' since 1980, in the distribution of planting material to farmers. Over the last six years arabica growers could obtain seedlings from the selected variety Java, which gives yields that are at least 50 % higher than traditional varieties, and has a higher resistance to coffee berry disease.

A major part of the present coffee plantations are too old and need to be replanted, but coffee production could be increased, in many other cases, merely through the application of better cultivation practices, such as the removal of shade, regular weeding and/or the establishment of cover crops, the use of fertilizers, spraying, etc. Extension services are provided by the Ministry of Agriculture and the development organizations in the respective areas (UCCAO, Zapi-Est, etc.). Theoretically, one extension officer from MINAGRI should assist about 250 farmers, but in reality he has to cover an area with at least 1500 farmers. In areas covered by the integrated rural development programmes, the extension activities are much more intensive. A major problem in the arabica zone is that coffee is generally interplanted with food crops and that the women, responsible for the food crop production, do not benefit from the coffee production and are therefore not particularly interested in the maintenance of the coffee trees.

8.7.2 Input supply and credit

In 1982/83 1.2 million cuttings and about 4 million seedlings were produced in central nurseries for the planting of robusta and arabica, respectively. These amounts are sufficient for 800 and 2600 ha, respectively. The beneficiaries did also obtain the necessary equipment and fertilizers. However the intension of the programme was to supply in 1982/83 planting material for 3000 ha robusta and 6000 ha arabica coffee. The actual distribution of planting material and equipment required funds to a total of CFAF 350 million, made available by Fonader. Planting material for robusta coffee is produced at only three centres, next to the research stations, and has to be transported over long distances to the respective production areas.

Other important inputs for coffee production are fertilizers, insecticides and fungicides. The Government, through Fonader, subsidized in 1982/83 a total amount of 85 000 tonnes of fertilizers, which constituted about 70 % of total quantities applied. The major were sulphate of ammonia and the compound fertilizers: 20-10-10, which were subsidized at a rate of 26 % and 50 %, respectively. Fertilizers are distributed by cooperatives, traders and in some areas by regional staff of the Ministry of Agriculture. A major part of these fertilizers were meant for the production of coffee and other export crops, but farmers and their wives are more inclined to apply fertilizers to their food crops. Although some fertilizers are produced in Cameroon, most of it is imported through Fonader and sometimes by cooperative organizations such as UCCAO.

In Cameroon, major pests and diseases for arabica coffee are the antestia and the coffee berry disease. Treatment against antestia with Lindane and recently also with Sumithion are undertaken every year by the Pest Control Unit, covering about 60 % of the total area (ha). Farmers themselves take care of the spraying with fungicides against coffee berry disease. For the treatment of some 85 000 ha, about 8 million packets of spraying material were distributed in 1982/83. But the amount of fungicides applied was in many zones still insufficient. The treatment against coffee berry borers on robusta plantations is both undertaken by the farmers themselves and, on a small scale, by aerial spraying. The farmers use the insecticides supplied (Dursban 4) also against ants.

External finance for coffee production is supplied by ONCPB, Fonader, Commercial Banks and the Central Bank (BEAC). Farmers can only obtain credit from Fonader, either on individual basis, if they possess a title-deed on their land, or as member of a cooperative (credit adhérent). But Fonader does not have the necessary regional infrastructure to act as an efficient credit organization for small farmers, and many of them depend for credit on traders, exporters and processing agencies. The marketing of coffee is mainly financed by the licensed exporters who from the start of the buying season, are able to obtain credit (up to a certain maximum related to their quotum) from the Central Bank, through their commercial banks. Storage constitutes a major cost element for which finance is required. Whereas coffee remains in better condition at high altitude, and own stocks can be used for speculation, exporters on the other hand benefit from a lower interest rate on their stocks if they deliver it soon to SOCOPAO for in transit storage.

8.8 Summary and economic parameters

Cameroon contains many different climates, ranging from a tropical rain forest climate in the south-west to semidesert conditons in the north. Good growing conditions for arabica coffee are confined to the highlands of West and North-West Provinces, whereas robusta can be grown in many areas in all but the northern provinces. The average yields of arabica coffee are low because of the old age of the trees, the intercropping with food crops, and general neglect. Robusta plantations, found in pure stand in less densely populated areas, give reasonable yields, even at low input levels.

The marketing channels in Cameroon vary not only by main type of coffee, but also by region. In some areas, such as West Province, cooperatives play a major role in the processing, marketing and even the export of coffee. The marketing of coffee in the other areas is regulated by the Produce Board, the ONCPB, and the processing and transport is undertaken by private traders.

ONCPB every year fixes the producer price and the respective marketing margins, and it uses the excess funds originating from the export of coffee, for price stabilization purposes and to finance development projects and financial institutions engaged in the improvement of the rural infrastructure. Is is questionable whether the country as a whole really benefits from this price stabilization programme: although the farmers continue to pick and deliver their coffee at the present low producer prices, they do not invest much in their plantations, a great many of which require rejuvenation or replanting.

Economic parameters of the coffee sector in Cameroon (1981/82).

Resources used in coffee sector	
Land under coffee	370 000 ha
as share of arable land	6 %
Labour involved	170 000 man-years
in production	130 000 man-years
in processing, trade & supporting services	40 000 man-years
as share of agricultural work force	4 %
Material inputs for coffee production	CFAF 8 000 million*
Materials used in processing & trade	CFAF 5 000 million*
Annual import requirement for coffee industry	US$18 million*
Income from coffee sector	
Value of coffee production	
(volume × export unit value f.o.b.)	CFAF82 000 million
as share of agricultural production	18 %
as share of GDP	5 %
Value of coffee exports (f.o.b.)	US$190 million
as share of agricultural exports	27 %
as share of exports	17 %
Share of total production value, obtained by	
Producers	46 %
Processing industries, cooperatives, traders & exporters	18 %
Government (through taxes, etc.)	36 %
Other parameters	
Robusta production	75 %
Arabica production	25 %
Yield (green coffee)	400 kg/ha (rob.), 170 kg/ha (arab.)
Production of labour involved (green coffee)	3.2 kg/man-day (rob.), 1.8 kg/man-day (arab.)
Share of coffee produced for export	98 %
Ratio producer price (per kg green coffee): daily wage rate (CFAF/man-day)	0.50 (rob.), 0.56 (arab.)

Sources: FAO (1971-1983), IMF (1983), World Bank (1984) and own estimates.

References

Bouharmont, P., 1982. La plantation du caféier Arabica au Cameroun. Café, Cacao, Thé 16 (1). Paris.

Bouharmont, P. & Awemo. J. 1979. La sélection végétative du caféier Robusta au Cameroun. Café, Cacao, Thé 23 (4). Paris.

CNCC, 1980a. Guide practique du café au Cameroun. Conseil National des Chargeurs du Cameroun, Douala.

CNCC, 1980b. Le café: Analyse du circuit d'acheminement en vue de l'amélioration de la compétitivité a l'exportation. Conseil National des Chargeurs du Cameroun, Douala.

Ediafric, 1983. L'agriculture camerounaise. Paris.
FAO, 1971-1981. Production Yearbook. FAO, Rome.
FAO, 1980-1983. Trade Yearbook. FAO, Rome.
FAO/World Bank Cooperative Programme, 1981. Report of the Cameroon Second Western Province Rural Development Project Preparation Mission. FAO, Rome.
GICAM, 1983. L'économie camerounaise. Bilan de l'exercise 1982-1983. Tome I. Yaoundé.
ICO, 1977-1983. Quarterly Statistical Bulletin on Coffee (several issues). ICO, London.
IMF, 1983. International financial statistics. IMF, Washington.
IRA, 1981. Rapport d'activités techniques. Yaoundé.
Les Atlas Jeune Afrique, 1980. Atlas of the United Republic of Cameroon. Editions J.A., Paris.
Ministère de l'Agriculture, 1981. Annuaire de statistiques agricoles, 1980-1981. Direction des études et projets, Yaoundé.
Ministère de l'Agriculture, 1983. Rapport annuel d'activités, campagne 1982-1983. Direction de l'Agriculture, Yaoundé.
Morel, Y., 1983. Tableaux économiques du Cameroun, 2eme édition. Collège Libermann, Douala.
ONCPB, 1983/84. Textes de campagne caféière 1982-1983 and 1983- 1984. Office National de Commercialisation des produits de base, Douala.
SEDES, 1982. Etudes des resources financières de coopératives café, cacao du Cameroun. Rapport définitif et annexes. Société d'Etudes pour le Développement Économique et Social, Paris.
UCCAO, 1983. Etudes sur les coûts d'exploitations des plantations paysannes de café Arabica et café Robusta. Bafoussam.
Université de Yaoundé, 1982. Revue de géographie du Cameroun 3 (2).
World Bank, 1980. United Republic of Cameroun Economic Memorandum. Washington.
World Bank, 1984. World development report, 1984. Oxford University Press, Washington.

Currency equivalents and price-indices in Cameroon (1965-1982).

	Exchange rate (CFAF per US$)	Consumer price index (1980 = 100)
1965	247	34
1970	274	37
1971	278	38
1972	265	41
1973	241	46
1974	230	53
1975	222	60
1976	225	67
1977	246	76
1978	226	86
1979	213	92
1980	211	100
1981	272	111
1982	329	126

Source: IMF (1983)

9 Coffee in Ivory Coast

9.1 Background

With its total area of 322 500 km², and with common borders with five other countries, Ivory Coast occupies a central position in West-Africa. This is among others reflected in the past and present migration patterns. Of the present total population of 8.9 million (1982) about a quarter constitutes of immigrants of recent date, in particular from Upper Volta. The country is divided into 8 regions, one of which is made up of the main urban centre, Abidjan, with about a million inhabitants. The regions are subdivided into 34 departments and 163 so-called 'sous-préfectures'.

The GNP per person was in 1982 about US$950, and the contribution of the respective sectors to the GDP was as follows: agriculture 26 %, industry 23 % and services 51 %. Of the total work force of about 4.3 million, about 79 % is working in agriculture, only 4 % in industry and 17 % in services (World Bank, 1984).

In the period to 1982 about 80 % of total exports consisted of agricultural and agro-industrial products, of which coffee with 20 %, cocoa with 30 %, wood and wood products with 12 %, and cotton and textiles with 3 % were the major commodities (FAO, 1980-1982).

Table 9.1 shows the major crops in Ivory Coast. With an average annual production of 292 000 t in the period 1979/1981, Ivory Coast contributed 6 % of the world production of coffee, and ranked third as world producer after Brazil and Colombia, and just before Indonesia. It is the first world producer of robusta coffee.

9.2 Ecology

Situated between 5° and 10° North of the Equator, and almost without high altitude zones, one finds in Ivory Coast everywhere a warm tropical climate with annual mean temperatures around 26 °C and throughout the year fluctuating between 24 and 30 °C only. With the exception of the northwestern mountain range with peaks above 1000 m, the altitude in the country is generally below 500 m (Figure 9.1).

The rainfall distribution throughout the year is mainly influenced by two airmasses: the hot, dry and dusty continental airmass and the humid Atlantic airmass. The border line between these two airmasses moves in a north-south direc-

Table 9.1. Average area harvested, yield and national production of major crops over 3-year periods.

Crops	Area (1000 ha)		Yield (kg/ha)		Production (1000 t)	
	1969/1971	1979/1981	1969/1971	1979/1981	1969/1971	1979/1981
Food crops						
maize	333	600	773	475	257	285
paddy	286	461	1168	1153	335	532
plantain*	700	800	933	1010	653	808
roots & tubers	555	834	4134	3830	2292	3194
Oil crops						
coconuts	.	.	.	.	49	155
groundnuts	52	63	800	846	42	53
oil palm[1]	68	101	.	.	65	183
Beverages						
cocoa	405	687	482	587	195	403
coffee	664	1060	366	275	243	292
Other crops						
cotton	40	119	909	1098	37	131
pineapple	.	300	.	1100	127	332
rubber	14	54	.	.	11	21
sugar-cane	.	19	.	5579	.	106

1. Including palm oil and palm kernels.
Source: FAO (1971-1983)

tion during the year. As a result, of them clearcut agro-ecological zones can be distinguished, with an average annual rainfall of about 2100 mm in the south and south-west to about 900 mm in the north-east (Figure 9.2). In between the tropical rain forest zone in the south and the savanna belt in the north) lies a large forest zone, including a transitional zone towards the savanna area, with an average annual rainfall between 1200-1700 mm, well spread over the year and this constitutes the coffee and cocoa area. Other major crops in this zone are yam, rice and bananas. The rainfall distribution is characterized by one dry season and a long rainy season with usually one peak in June and another peak in September (Table 9.2).

Although the total annual rainfall is on average low for cultivation of robusta coffee, the rainfall distribution and temperatures are satisfactory. Temperatures are generally too high for arabica coffee. Major crops in the southern rain forest zone are oil palm, bananas, cassava and pineapple, and in the northern savanna area cotton, rice and other grains.

The red and porous ferrallitic soils or oxisols (ferralsols and acrisols in the FAO/UNESCO-system) are the most common soils in Ivory Coast. In the high rainfall areas in the south, the east and the central zone in the country these are

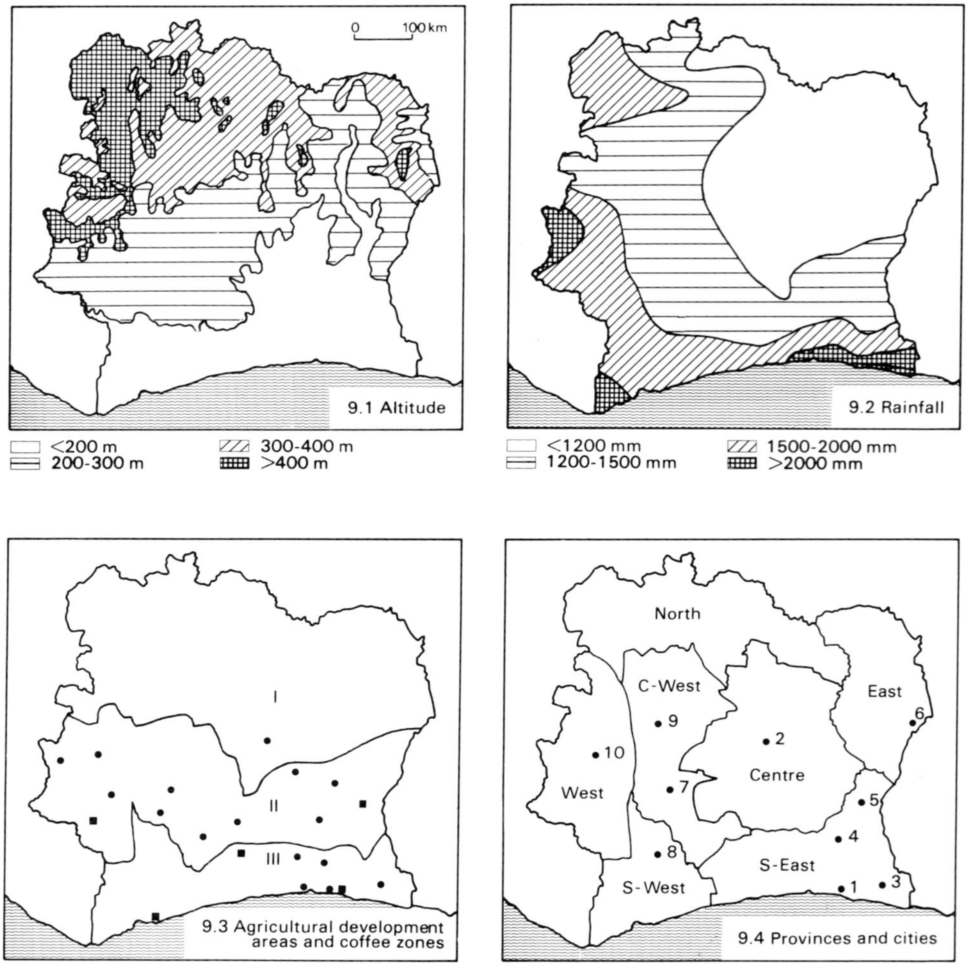

Fig. 9.1-9.4. Altitude (9.1), average annual rainfall (9.2), agricultural development areas and coffee zones (9.3) and provinces and major cities (9.4) of Ivory Coast. After Les atlas jeune afrique (1983).

Table 9.2. Distribution of rainfall by region.

Region	Station	Annual rainfall (mm)	Monthly rainfall (mm) J	F	M	A	M	J
East	Bondoukou	1162	13	40	84	131	164	160
S. East	Adzopé	1428	19	52	112	160	197	264
Centre	Bouaké	1178	12	43	92	136	143	146
C. West	Séguéla	1338	16	41	93	115	144	133
West	Man	1710	16	56	114	159	158	199
S. West	Soubré	1524	30	61	128	151	171	222
			J	A	S	O	N	D
	Bondoukou		84	72	185	170	44	15
	Adzopé		159	60	122	170	83	30
	Bouaké		103	107	207	134	35	20
	Séguéla		120	174	248	175	55	24
	Man		202	261	314	158	52	21
	Soubré		114	97	211	175	113	51

Source: Ministère de l'Agriculture (1981)

strongly weathered, and the chemical fertility is generally low and the layer of top soil thin. But the soil structure and soil depth are adequate for crops like coffee and cocoa.

9.3 The coffee sector

9.3.1 History

The first commercial coffee plantations were established in Ivory Coast around 1880 by the Frenchman Verdier, in the Aboisso area. This concerned varieties of the species *Coffea canephora*; var. Kouilou (robusta) and *C. liberica*. In the south-east the development of the cocoa and coffee plantations took place as fast as in neighbouring Ghana. But in the central and western part of the country the colonial government intervened after 1925 to stimulate the plantation economy. The local population originally showed little interest in the establishment of tree crops. They were forced to work on the European plantations, and did not receive any incentives to develop their own plantations.

After World War II a major expansion of agricultural activities took place. This was made possible through the extension of the network of roads and the opening up in 1950 of the Vridi-canal, which provided Abidjan with an excellent sea harbour. The colonial government, realizing the potential of African smallholder plantations, then abolished the obligatory work regulations and started to distribute on a large scale planting material[1] and other inputs.

In the past local leaders and heads of family arranged for the distribution among their people of the communal land, which was cultivated with food crops following shifting cultivation patterns. Whereas originally the women were mainly involved in cultivation activities, with the disparition of hunting the men had gradually become more interested in agriculture. And once the opportunity was given to establish permanent crops the men with their nucleair families saw this as a means of obtaining some independence within the clan, and the clearing of forest-land since then took place at an alarming rate. The supplementary labour force for the plantation economy was attracted from the northern savanna zone and subsequently from neighbouring countries.

To protect the interests of the small planters, but also to serve as an instrument to control the national production, the 'Caisse de Stabilization des Cours' was established in 1955. And in 1959 more than 200 000 farm families produced coffee and cocoa, against only 40 000 families in 1944.

9.3.2 Recent developments

From the above it follows that the areas planted under coffee and cocoa increased rapidly since 1950. In 1973 more than a million hectares were planted with coffee and 0.7 million hectares with cocoa. And in 1981 the area for coffee was 1.2 million ha (of which 1.05 million adult plantations) and for cocoa 1.1 million ha (0.8 million adult plantations). Whereas the average yield for cocoa seems to have increased gradually over the years, the average yield for coffee not only fluctuated heavily, but did not show any upward trend. Through the faster growth of area and yield the national cocoa-production surpassed that of coffee in the late 1970s (Table 9.3).

Coffee and cocoa are grown throughout the forest zone between 5° and 8° North of the Equator, but the coffee production was concentrated particularly in the southern and central regions. In the 1970s however, the coffee production declined in the central region (which is in fact less suitable for coffee) and increased in the western regions, where enough land was still available. Although the national average yield is extremely poor, reasonable yields were, in the period 1977-1980, obtained in the southern and central western regions (Table 9.4). The national coffee and cocoa development organization, SATMACI, has through its nurseries, which issue free of charge planting material of robusta hybrids to farmers, succeeded to add in the period 1972-1983 about 100 000 ha of reasonably high yielding plantations (600-800 kg/ha). The organization has also begun to provide

---

1. In the period 1945-1948, a fungal disease attacked a large proportion of the coffee plantations and the new plantations therefore consisted entirely of the new Robusta variety, introduced from Congo (Zaire).

Table 9.3. Area planted, average yield and national production of green coffee and cocoa beans (1950-1981).

Crop year	Coffee			Cocoa		
	area (1000 ha)	yield (kg/ha)	production (1000 t)	area (1000 ha)	yield (kg/ha)	production (1000 t)
1950/51	192	272	52	167	341	57
1955/56	343	262	90	207	324	67
1960/61	396	468	185	261	358	93
1965/66	615	443	272	327	346	113
1970/71	674	356	240	404	443	179
1975/76	901	342	308	498	464	231
1976/77	921	316	291	526	442	232
1977/78	951	206	196	557	569	304
1978/79	1010	274	277	586	543	318
1979/80	1046	239	250	752	533	401
1980/81	1064	345	367	777	537	417
1981/82	.	.	248	.	.	.

1. The area under cocoa and average yields could well be under-respectively over-estimated.
Sources: Ministère de l'Agriculture (1981), Bonnefond (1980)

Table 9.4. Area planted, average yield and national production of coffee by region over 3 year periods.

Region	Production (1000 t)		Area 1977/1980 (1000 ha)	Yield 1977/1980 (kg/ha)
	1968/1971	1977/1980		
East	5.8	10.6	58	183
South	75.8	83.5	288	290[1]
Centre	86.4	50.6	278	182
C. West	38.4	48.7	165	295
West	33.3	39.6	177	224
S. West	2.9	8.8	37	238
Total	242.6	241.8	1003	241

1. Highest yield in Aboisso (400-600 kg/ha).
Source: Ministère de l'Agriculture (1981)

subsidies to farmers to rejuvenate their plantations. This is an important activity, since a large proportion of plantations is now older than 20-25 years. A major achievement of the coffee and cocoa research organization, IRCC, has been the development of the Arabusta variety (an interspecific hybrid between *C. arabica* and *C. canephora*). The further introduction of them depends on the possibility of obtaining a separate position on the world market and on the introduction to farmers and their groups of improved production and processing methods, essential for Arabusta.

9.3.3 *The organization*

Coffee is produced by a limited number of large farms and by about 350 000 smallholders with an average size of plantations of about 3 ha (1983). Whereas in the past coffee was dried, hulled and sorted at the village level, since the establishment in the period 1978-1980 of 16 large central hulling factories, dried cherries are collected from the villages and taken to the factories by private buyers. After hulling, transporters deliver the coffee to the port facilities of the exporters, where the coffee is graded and bagged for export. Only about 15 000 tonnes per year are roasted and subsequently consumed locally. A small amount is processed into instant coffee.

Technical assistance and extension services to cocoa and coffee-farmers are provided by SATMACI. This organization, created in 1958, is the oldest of the state-owned agricultural development companies, and is now more or less responsible for the agricultural development in the whole forest or coffee and cocoa-zone. SATMACI employs a total of about 1100 people. Similarly SODEPALM and CIDT are responsible for agricultural development activities in the southern

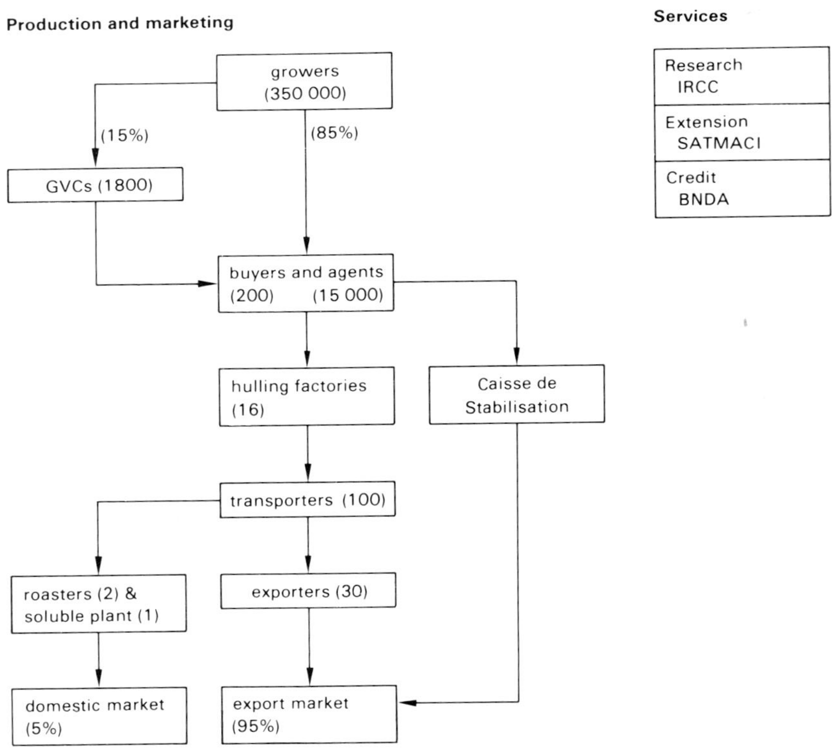

Fig. 9.5. Participants in the coffee sector and main marketing channels in Ivory Coast in 1982.

and the northern belt, respectively. These development companies are also engaged in input supply, in the commissioning or construction of rural infrastructural works and in the management of loans.

The marketing of coffee and cocoa is, however, controlled by another parastatal organization: the 'Caisse de Stabilization' (CSSPPA). This organization is legally the owner of all marketed coffee and it establishes quaranteed prices to producers of cocoa, coffee, cotton and oil crops (groundnuts), inspects the quality of these products and controls transport, storage and export sales. Through pricing policies and accounting procedures for the respective operations, large financial surpluses occur during periods of high world-market prices, which are used to build reserves for re-investment in agriculture and rural development and partly for the Government's investment programmes in other sectors. The total capital stock of the Caisse is about CFAF 4 thousand million (1982) and the organization employs about 1400 people. It also has offices in New York, London and Paris.

Research on coffee and cocoa is undertaken by the IRCC, which acts also as a Technical Advisor to SATMACI. Agricultural credit is mainly supplied through the BNDA (Banque Nationale pour le Developpement Agricole). Figure 9.5 gives an overview of the participants in the coffee sector and the main marketing channels for coffee in Ivory Coast. The expansion of coffee production among smallholders until the late 1960s can be attributed to initiatives by Government and private organizations. Thereafter, farmers needed no further encouragement to start or expand coffee plantations. Now the Government regulates coffee production through its internal price (stabilization) policies and the recent coffee rejuvenation programme. These measures appear, however, to be insufficient to redress the recent trend of a coffee production lagging behind that of cocoa.

9.4 Production

9.4.1 Production units and farming systems

Over the past twenty years a rapid growth of agricultural production took place, particularly in the forestzone, and this is reflected in the changes that the agricultural production units underwent. In Table 9.5 some parameters of agricultural production are given for 1975 as well as the changes thereof, over the period 1965-1975.

The table shows that during the period 1965-1975 the agricultural work force and the area cultivated in these areas increased at an average annual rate of 6 % and 7 %. These increases can be largely attributed to the influx of immigrant farmers and seasonal labourers, particularly from Upper-Volta. The total value added increased at a similar or slightly lower rate, resulting in 0 % and 1 % annual growth rates for the value added per labourer and per hectare, respectively. Because of the increasing share in coffee and cocoa in total production. the cash earnings per labourer increased, however, at about 3 % per year.

Table 9.5. Structural and economic parameters of agricultural production in the major coffee and cocoa production zones (1975). In brackets proportion of annual growth (%) in period 1965-1975.

	Agricultural zone (and main town)				
	South East (Abengourou)	Central West (Daloa)	Centre (Bouaké)	West (Man)	Total
Agric. work force (× 1000)	533 (8)	520 (7)	264 (2)	225 (2)	1542 (6)
Number of farms (× 1000)	116 (4)	134 (5)	83 (1)	70 (1)	403 (3)
Area cultivated (× 1000 ha)	820 (4)	821 (11)	219 (3)	308 (9)	2168 (7)
Av. work force per farm	4.6 (4)	3.9 (3)	3.2 (1)	3.2 (1)	3.8 (3)
Av. farm size (ha)	7.1 (0)	6.2 (7)	2.6 (1)	4.4 (8)	5.4 (4)
Value added per labourer (CFAF 1000)	108 (−4)	115 (+2)	96 (−1)	81 (4)	104 (0)
Cash earnings per labourer (CFAF 1000)	81 (−2)	108 (+7)	43 (0)	41 (+5)	78 (+3)

Source: de la Vaissière (1981)

There are, however, large differences between the four major production zones. The south eastern zone in 1975 still had the highest average farm size (7.1 ha) and the highest average work force per farm (4.6), but the area cultivated and the average farm size were rapidly increasing in the centre-west and western zone (de la Vaissière, 1981). The demographic and economic stagnation in the central zone can partly be explained by the establishment of the Kossou-dam and its reservoir, but also by more general migration patterns towards the centre-west zone (World Bank, 1982).

In the forest zone, production units can be classified on the basis of the presence, next to food crops, of either coffee, cocoa or a combination of those crops. In the late 1970s about 80 % of the farms in the forest zone had some coffee and nearly 60 % some cocoa (Table 9.6). Only 15 % of the farms had no tree crops at all. This concerned small farms with hardly any paid labour. Almost 35 % of the farms had only coffee trees and food crops (e.g. yams, cassave, plantain, paddy, maize). Many of those were in the category of small to medium-sized farms with a permanent work force of less than 5 people and relying much on seasonal labourers[2]. More than 40 % of the farms had both coffee and cocoa plantations and more than a third of those could be considered large farms (work force of 5 or

2. Paid labour can be classified as
– permanent labourers, often provided with housing and a plot for production of food crops
– sharecroppers
– labourers on piece-work contract
– seasonal labourers, hired on a daily basis.

Table 9.6. Proportional distribution (%) of size of holdings in the forest zone (Southern Ivory Coast) (1974/75).

Size of holding (ha)	Coffee & cocoa	Coffee only	Cocoa only	Neither coffee nor cocoa	All holdings
less than 1	1	5	6	42	9
1-2	7	16	16	33	15
2-5	32	49	43	20	36
5-10	38	24	23	3	27
more than 10	22	6	12	2	13
Number of holdings (× 1000)	199	149	26	69	443

Source: Lee (1980)

more). For small farms coffee and cocoa are often competitive crops.

The distribution of production factors by farm appeared to be rather uneven, with such factors as age of farmer, number of wives, family size, farm size, number of enterprises, number of paid labourers, etc., being often closely related. This is undoubtedly also reflected in the distribution of income of households. The farms of immigrants were more oriented towards cultivation of cashcrops and did expand more rapidly, than those of the autochtone population. Whereas the women are generally responsible for the food crop production, the men are more involved in cultivation (harvesting) and marketing of the tree crops.

Whereas coffee is generally interplanted with food crops in the first two years after establishment, about 60 % of adult plantations are in pure stand. Of the mixed stand plantations about 25 % are with cocoa and about 15 % with colatrees. Not less than 28 % of the present plantations date from before independence and 33 % are between 15-25 years old. In other words, a large proportion of all coffee plantations needs to be rejuvenated or replanted. Only 8 % of the present (1982) plantations is younger than 5 years.

9.4.2 *Production stages and production costs*

Although the rejuvenation of old plantations has now been initiated, the large majority of the present coffee and cocoa plantations has been established after the clearing of virgin forest bushland. This involves the cutting and removal of trees and strubs, followed by burning and cleaning. Meanwhile farmers establish their nursery. In the past multiplication took place by seed, but now farmers can obtain rooted leafcuttings from selected robusta-clones, produced by some 35 multiplication centers (centres de bouturage) of SATMACI. For each hectare to be established farmers can obtain 1750 cuttings and 1500 polyethylene bags (250 cuttings extra for replacements). In 1983 the costs (borne by SATMACI) amounted to CFAF 22 per cutting and CFAF 7 per bag. The coffee plants remain for about 9

Table 9.7. Establishment costs (years 1-5) per hectare for coffee and cocoa plantations (1982/83).

Year	Activities	Coffee		Cocoa	
		traditional	modern	traditional	modern
1	Clearing, burning, nursery, layout & digging (man-days)	40	40	40	40
2	Planting & weeding (man-days)	90	180	58	122
	Material costs (CFAF 1000)	12	48[1]	7	20[1]
3/4/5	Maintenance (man-days)	55	80	40	60
	Material costs (CFAF 1000)	–	20	–	12
	Total establishment costs (CFAF 1000)[2]	169	323	124	221

1. Largely paid by SATMACI.
2. Average wagerate of CFAF 850 per man-day.
Source: Personal communications

months to one year in the farmers' nurseries before being planted in the field, usually in the period April to June. Planting holes should be about 40 cm^3.

Table 9.7 shows the total costs of establishment of a modern coffee plantation, which were in 1982/83 about CFAF 320 000 per hectare, or twice as much as for a traditional plantation (with a lower planting density and multiplication by seed). The establishment costs for cocoa plantations are about 30 % lower than for coffee.

Table 9.8 presents the annual costs of production of coffee and cocoa for different type of plantations, mainly related to the age of the plantations. Old plantations generally receive little attention, and weeding operations are only undertaken once or twice. That could be sufficient for cocoa plantations, since those are usually found under shade, but not for coffee trees, that are generally cultivated without shade. Fertilizers are seldom applied, whereas spraying with pesticides and fungicides is only undertaken by advanced smallholders on high yielding plantations. Some smallholders used to apply HCH against ants.

Harvesting is done according to the stripping method. The harvested cherries are spread out on sheets, banana leaves etc. or on drying tables for a period of about 10 days, to dry in the sun. More and more use is made now of plastic sheets to protect the coffee against the rain. Cocoa requires some more postharvest processing, since the beans have first to be removed from the pod.

Table 9.8 partly explains why farmers generally favour the cocoa plantations in recent years and why average yields of cocoa are now above those of coffee. The expansion of coffee plantations took place before that of cocoa, and a relatively larger proportion of coffee plantations is now older than 20 years. Whereas the average cost of production of cocoa was in 1982/83 in the order of CFAF 170 per kilogram dried beans, or just over half of the producer price in that year, the aver-

Table 9.8. Production costs (per hectare and per kilogram) of coffee and cocoa for different types of plantation (1982/83).

	Coffee			Cocoa		
	traditional		encadrée[2]	traditional		encadrée[2]
Age of plantation (year)	20	10-15	10	20	10-15	10
State of maintenance	bad	reasonable	good	bad	reasonable	good
Yield[1] (kg/ha)	150	350	550	180	500	650
Plant density (plants/ha)	700	950	1400	400	800	1200
Wage rate (CFAF/day)	850	850	850	850	850	850
Annuity of establishment costs (CFAF 1000) (10 %, 20 years)	–	20	38[3]	–	15	26[3]
Annual costs						
Labour inputs (man-days)						
weeding	26	30	34	14	20	20
disease control	–	–	2	–	4	4
pruning	–	–	3	–	–	2
harvesting	20	38	54	8	17	20
processing	–	–	–	6	14	18
drying	1	2	3	4	6	8
transport	1	2	3	1	3	4
other	2	3	3	2	3	3
total	50	75	102	35	67	79
Subtotal of costs (CFAF 1000)	43	64	87	30	57	67
Material inputs (CFAF 1000)						
insecticides & fungicides	–	–	2	–	7	7
processing & other	2	3	5	2	4	5
Subtotal	2	3	7	2	11	12
Total costs per hectare (CFAF 1000)	45	87	132	32	83	105
Costs per kg (CFAF)	300	249	240	178	166	162
(US$)	0.91	0.76	0.73	0.54	0.50	0.49

1. Respectively green coffee equivalent and dried cocoa beans.
2. Established under SATMACI programme.
3. Including planting material obtained from SATMACI.

Source: Adapted from Ruf (1980d)

age costs for coffee were about CFAF 140 per kilogram cherries (US$0.40/kg) or just below the producer price (CFAF 300/kg for green coffee, converted into a payment of CFAF 150/kg for cherries).

Table 9.9 illustrates that coffee and cocoa are competitive crops for the major production factor labour, which is generally needed in about the same periods in the year.

Table 9.9. Distribution of cultivation activities for coffee and cocoa throughout the years.

Months:	/	J	F	M	A	M	J	J	A	S	O	N	D	/	J
Coffee:			weeding	pruning		weeding			weeding			harvesting/drying			
Cocoa:		spraying			(weeding)			weeding				harvest/drying/processing			

9.4.3 *Yields and returns*

Although the average annual yields of coffee fluctuate considerably through a combination of factors such as weather conditions, biennial cycle, etc., there is now clear evidence of an gradual declining production per hectare of coffee. Whereas the average annual yield reached about 370 kg/ha in the 1960s, this has dropped to about 270 kg/ha in the last five years. This is largely due to the excessive age of most coffee plantations, which need to be rejuvenated.

However it is also possible that an increasing part of coffee plantations is no longer maintained nor harvested, because of the present price relationship between coffee and cocoa, and between the producer price for coffee and the wage rate for agricultural labour (Ruf, 1980b). Table 9.8 shows that for old plantations the production costs per kilogram were in 1982/83 about equal to the producer price per kilogram. The average return to labour was exactly CFAF 850 per man-day, and no remuneration was left for farmers using only paid labour. For traditional plantations (10-15 years) and for modern plantations (less than 10 years old) the average return to labour would, according to these calculations, have been about CFAF 1100 and CFAF 1200 per man-day. The average return to labour for cocoa plantations on the other hand, ranged between CFAF 1500 and CFAF 2000 per man-day. For the 1983/84 season the producer price for coffee was increased to CFAF 350 per kilogram, but the cocoa price was increased proportionally.

The extensive methods of coffee and cocoa-cultivation, using much land and little labour, can be understood from the farmer's point of view, but it can be questioned whether the disappearance of large forest reserves are justified from the national viewpoint.

9.5 Processing

The processing of coffee, from cherries into green coffee, was until 1978 carried out entirely at the village level. The cherries were dried and sorted by the farmers, and subsequently taken to a trader, sometimes a farmer himself, who owned a small grinding mill, that was adjusted in such a manner as to be able to hull the dried cherries. Farmers were only paid for adequately dried, well sorted cherries that after hulling produced good quality green coffee. Black beans were removed.

Many of these installations were, however, not operating adequately; the turnout was often less than 50 % and the proportion of broken beans high. It was therefore decided to establish a certain number of central hulling factories in the respective production zones.

The first and until now the largest plant was established in Toumbokro and started its operations in 1973/74. Plans were made to built 8 more plants of the same size, but eventually 15 other (smaller) plants were built in the period 1978 to 1980 (Table 9.9). With these factories, the average conversion factor green coffee: dried cherries increased from about 47 % to 53 %, mainly because of the considerable reduction of the proportion broken beans. The factories are normally in operation for about four months (December to March), which in years of a peak supply could be extended to five months. Several factories are also engaged in the supplementary drying of cocoabeans.

The total investment in these processing plants amounted to about CFAF 30 thousand million, and the plants together employ approximately 800 permanent staff and over 2000 seasonal workers during the processing period. The annual costs of operation is about CFAF 300 million for an average sized plant, with fixed costs representing about 75-80 %. Variable costs consist mainly of the costs of energy, seasonal labour and spare parts.

Traders (traitants) and their agents now collect the dried coffee cherries from the farmers and bring these to the factory. The processing includes among others the removal of the skin (dried pulp) and the actual hulling. In some cases supplementary drying is required (to 13 % humidity or less). Only a few factories are also involved in the classification. This is usually done at the premises of the exporters close to the port.

This classification (grading and sorting) includes several steps, starting with the elimination of stones, dust, etc, followed by the rejection of black beans by electronic (colour) sorting equipment and subsequently followed by grading (calibrage). Apart from the 4 regular export grades, elephant beans and broken beans are distinguished. The latter are grounded and used for local consumption. The quality can be 'courant' or 'superieur'. In the end the coffee is bagged (at a speed of 200 bags per hour) and the bags are piled up in stacks ('lots') of 420 bags, three of which contain together about 75 t.

Roasting is only undertaken by three companies, one of which is a subsidiary of Nestlé and produces soluble (spray-dried) coffee, 92 % of which is exported (in particular to Greece and some Arab countries). The plant has an annual production of about 6000 t instant coffee and provides employment for approximately 400 people. The Caisse allows the plant to buy the raw material (green coffee) at the local integral costprice, instead of the world market price.

9.6 Marketing

9.6.1 *Marketing channels*

The dried cherries are collected from the farm or from the centre of the village cooperative by one of the about 15 000 agents, that operate for the approximately 200 authorized private buyers (including exporters). The buyers are nominated by a regional committee and licensed by the Caisse. A bonus ('prime de groupage') is paid to farmers who bring the cherries to the centre of their GVC (Groupement à Vocation Coopérative).

The Caisse de Stabilization exercises control over the collection of cherries. A disadvantage of the central processing system is that the quality delivered by farmers cannot be detected easily on the spot. Some experiments are undertaken to use mini-hulling equipment in the villages to check upon the quality of the dried cherries.

The transport of green coffee from the sixteen hulling factories to the warehouses of the exporters (about 30 in total) is undertaken by specialized transporters. Some of these have recently, in conjunction with exporters, introduced bulk transport in containers, which eliminates the labour-intensive loading and unloading of bags. The main port of export is Abidjan, but some coffee is exported from San Pédro.

9.6.2 *Marketing functions*

During the period of collection, processing and marketing of coffee (from November to April), which coincides largely with that for cocoa (October to February), a considerable proportion of the national fleet of lorries and pick-ups is tied up with the transport of coffee and cocoa. The capacity of these vehicles ranges from a mere 1 t for pickups to not less than 30 t for the largest lorries. In the collection of coffee cherries alone an estimated number of 1500 lorries and 4000 pickups are involved. Transport of hulled coffee from the factories to the port facilities is undertaken by Union affiliated transporters at a rate, fixed by the Ministry of Transport, of CFAF 29 (US$0.07), per tonne-kilometre (1983/84).

Whereas all authorized transporters are required to have their own stores, the largest warehouses for the storage of coffee are owned by the exporters and the processing firms. The hulling factories have all their own stores for cherry as well as green coffee, and these are sometimes also used for cocoabeans. Cocoa generally requires more space, because it cannot be piled up as high as coffee (risk of overfermentation). Now there are about 35 approved warehouses for green coffee in the country with a total capacity of more than 10 million 60-kilo bags. Most of these are in Abidjan. Total held-over stocks of exportable coffee, verified every year at 30 September, fluctuated in the past 7 years between 0.6 and 3.4 million bags, whereas average annual exports were around 4.2 million bags.

9.6.3 *Marketing margins and taxations*

Although the Caisse is not involved itself in the physical handling of coffee and some other products, it is responsible for the marketing of these products from the farmgate to the export market, and part of the exports are made on its account. At the beginning of each crop year the Government publishes the minimum price payable to growers for these crops. Subsequently the Caisse lays down the various charges, that the buyers, the hulling factories, the transporters and the exporters, can make for their services between the point of purchase and the final shipment. These costs and margins are together with the taxation rates presented in the so-called 'Différentiel-café' for that particular crop year (Table 9.10).

After export the Caisse settles account with the exporters who in turn pay the agents and transporters. If the Caisse ends the season with a positive balance, 60 % of this sum is credited to the Price Stabilization Reserve, 30 % is used for the improvement of the rural infrastructure and the remaining 10 % for agricultural credit schemes. One clear disadvantage of relative stable internal prices is formed by the illicit border trade with countries where the producer prices do follow more closely the prices on the world market.

Table 9.10 gives an overview of the respective cost elements in the processing and marketing of coffee for two periods: 1976/77 and 1983/84, and for the first period for both the traditional village level and the centralized industrial processing (hulling and sorting). In the calculations for the period 1976/77 it was assumed that the hulling operation would give in both cases an outturn of 55 %. But because of a large proportion of broken beans, etc., the village level processing would show a final conversion factor green beans: dried cherries of 0.47 against 0.52 for centralized industrial processing. According to these calculations the centralization did not reduce markedly the unit costs of processing and marketing. However the scheduled figures for 1983/84 suggest that the efficiency increased in recent years since the ratio payment to producer : integral costprice f.o.b. is estimated at 68 % against 60 % for 1976/77.

Table 9.10. Hulling factories and their annual processing capacities (1000 t green coffee).

Firm	Locations	Capacity
SERIC	Toumbokro	35
Unicafé	Daoukro, Divo, Issia, Man	80 (20 each)
	Gagnoa, Anyama	30 (15 each)
	Aboisso	30
UTPA	Abengourou, Daloa, Kotobi	90 (30 each)
Decorticaf	Oume, Danone, Duekoue	45 (15 each)
CIPRO	Sikensi	12
SHAD	San-Pedro	10
Total	16 factories	332

To estimate the share in the average unit value of exports, obtained by producers, these unit values are also indicated in Tabel 9.10, but the period 1976/77 has been exceptional with steep rising export prices. In Table 9.11 the ratio between the unit value of exports and the producer price is given for the whole period 1975-1982. Over this period farmers received on average about 40 % of the unit value of exports, which is low compared with other major coffee-producing countries.

Table 9.11. Prices, marketing costs and margins in CFAF and US$ per kilogram (green coffee equivalents) in the crop years 1976/77 and 1983/84.

	Village level/Central Processing 1976/77 (CFAF/kg)		Scheduled for 1983/84 (CFAF/kg)	(US$/kg)	Share of 1983/84 export unit value (f.o.b.) (%)
Payment to producer[1]	175	174	330	80	40
Costs of collection[1] (and transport)					
transport & storage			25	6	
bags			2	0	
margin buyer			3	1	
Total	15	16	30	7	4
Costs of processing & curing					
hulling	12		29	7	
sorting	6	21	5	1	
grading	2		5	1	
bagging	5	5	8	2	
interest	2	2	4	1	
general costs	6	6	4	1	
taxes (T.P.S.)[2]			1	0	
Total	33	34	56	14	7
Costs of exportation					
transport, storage	4	4	3	1	
port-handling	2	2	3	1	
margin to exporter	2	2	3	1	
customs (export-taxes)	59	59	57	14	
taxes (T.P.S.)[2]			2	0	
Total	67	67	68	17	8
Integral cost price	290	291	484	118	59
Average export unit value (f.o.b.)	increasing from 414 (1976) to 1006 (1977)		820	200	100

1. Conversion to hulled/green respectively: 55/47 %; 55/52 %; 55/53%.
2. T.P.S. = 'Taxe Présentation de Service'; rate = 25 % × 1.0153.
Sources: SERIC (1977), CSSPPA (1984)

9.6.4 *Export and domestic consumption*

About 95 % of the coffee produced in Ivory Coast is exported, and more than 97 % of these exports are in the form of green coffee. Another 2.5 % is exported as soluble and a mere 0.1 % as roasted coffee. Efforts are undertaken to boost the domestic consumption, which was largely made up of the lower grades robusta coffee. The annual consumption per person of coffee is about 1.6 kg, which is low compared with other major coffee-producing countries, but high compared with other African countries.

Table 9.12 gives the total production, consumption and exports over the period 1974-1983. Of total exports about 86 % is shipped to ICO-member countries, the most prominent among which are France, the United States, Italy and the Nether-

Table 9.12. Share of the export unit value obtained by the producer of coffee (1975-1982).

	Total value of exports (f.o.b.) (US$ 10^6)	Export unit value		Price to producer (CFAF/kg)	Share of export unit value to producers (%)
		US$/kg)	(CFAF/kg)		
1975	320	1.30	313	151	48
1976	642	1.93	414	158	38
1977	1017	4.21	1006	198	20
1978	783	3.19	784	251	32
1979	893	3.22	685	265	38
1980	731	3.38	714	302	42
1981	517	2.17	590	303	51
1982	511	1.89	621	303	49

Source: ICO (1977-1983)

Table 9.13. National production, domestic consumption and export (in 1000 t) (1974-1982).

Crop year	Opening stock	Pro-duc-tion	Con-sump-tion	Available for export	Export		
					total	ICO members	non-members
1974/75	53	270	3	320	214	175	39
1975/76	58	308	9	357	332	281	51
1976/77	32	291	4	319	276	256	21
1977/78	43	196	11	228	229	196	33
1978/79	−1	277	13	263	277	240	37
1979/80	3	250	13	240	194	157	37
1980/81	43	367	13	397	215	180	35
1981/82	186	248	13	421	284	244	40
1982/83	138	280	13	405	268	231	37

Sources: Ministère de l'Agriculture (1981), ICO (1977-1983)

Table 9.14. Proportional distribution (%) of coffee-exports to major importing countries (1972-1982).

	Four years periods		Year
	1972/73-1975/76	1977/78-1980/81	1981/82
France	33	25	22
United States	20	14	18
Italy	4	10	13
Netherlands	8	15	10
United Kingdom	3	6	7
Algeria	6	8	6
Japan	5	5	4
Spain	3	4	4
others	18	13	16

lands. The major non-ICO-member country to which Ivory Coast exports coffee is Algeria. The proportional distribution of coffee-exports to the respective importing countries is shown in Table 9.13 for three periods.

9.7 Supporting services

9.7.1 Research

Research on coffee is carried out by IRCC (Institut de Recherches du Café et du Cacoa et autres plantes stimulantes), which has five statons in different parts of the country, the major of which is in Divo. IRCC-headquarters are in Bingerville, not far from Abidjan. Research activities for both robusta and arabusta coffee are divided into four major areas: genetic improvement, agronomy, pest and disease control, and technology.

Major criteria for the selection of new robusta clones, to be multiplied and distributed through SATMACI, are: level of yields, beansize and caffein content. With recent mixtures of clones yields were obtained, at the research stations, of 2.5 tonnes green coffee per hectare, with an average caffein content of 2.7 %. Much emphasis is laid on a further reduction of the caffein content (IRCC, 1982).

The agronomic research into the possibilities of intercropping of food crops has led to the conclusion that this could be recommended, under certain conditions, for rice, groundnuts and yam. Bananas and taros would give excessive shade under the climatic conditions of Ivory Coast and maize would too much compete with coffee for nutrients. Important trials have been carried out on fertilization and irrigation of coffee, and research was undertaken on the biology and ecology of a locust *Zonocerus variegatus*, swarms of which represent a major threat to coffee plantations.

IRCC is also involved in the analysis of samples of cherries and green coffee, from the hulling factories, to examine the efficiency of coffee processing.

Since 1962 IRCC is involved in the production of the interspecific hybrid arabusta (*C. canephora* × *C. arabica*), the first families of which were planted in 1967/68. Since then, experiments have been carried out with a great number of arabica cultivars (from Central South America and Ethiopia) and robusta plants, and out of the arabustas obtained, a mixture of ten clones has been selected, which on average yield about 1.5 tonnes green coffee per hectare (at the research station). The caffein content is about 1.5 %. Agronomic experiments showed that optimum planting density is at any rate above 2000 plants per hectare. The newly created organization CEDAR established in 1975 an industrial coffee plantation of 350 ha near Soubré to carry out practical experiments on a larger scale. Arabusta seems to be vulnerable to antestia, but coffee rust does not harm arabusta.

In quality arabusta finds itself in between arabica and robusta, but similar to arabica it requires the method of wet processing and it needs careful cultivation practices. It is therefore envisaged to introduce arabusta coffee to groups of 50 small farmers, by means of block-plantations of 50 ha, the first of which will be established in Issia, Sassandra and Soubré in hitherto uncultivated forest zones.

General agronomical and ecological studies are undertaken by ORSTOM which organization is also involved in a coffee breeding programme in collaboration with IRCC. Agro-economic studies are carried out by CIRES, GERDAT and ORSTOM's centre at Petit Bassam. The Ministry of Agriculture has itself a department BETPA responsible among others for feasibility studies and for monitoring the processing and export of coffee.

9.7.2 *Extension services*

SATMACI is responsible for all extension activities related to coffee production. It has in the coffee and cocoa Zone 6 regional offices and 22 district centres, and employs more than 900 extension agents. They advise farmers on such matters as the establishment of new plantations, cultivation practices and disease control and particularly the possibilities of rejuvenation and replanting of coffee.

In recent years more than 10 000 ha were each year planted with robusta hybrids and about 7000 ha coffee were every year rejuvenated. These areas are to be compared with the about 40 000 ha that become unproductive every year. To maintain national production levels the yields of these new and renewed plantations should be at least two to three times as high as those of traditional plantings. The planting material for the hybrid plantations consists of cuttings which are produced at 35 so-called 'centres de bouturages', that have at their disposal about 20 000 coffeeplants from 8 clones, on a plot of 3 to 4 ha. These centres can produce 3 times per year about half a million cuttings, of which about 60 % root successfully. On average, a centre employs about 20 people (man-years).

SATMACI also manages some industrial coffee plantations, and it operates two training centres for three monthly courses on coffee and cocoa-production.

The theoretical aspects are dealt with at the centre in Gagnoa and practical work is done at the centre in Divo.

9.7.3 *Credit and input supply*

Agricultural credit is formally provided by the BNDA (Banque Nationale pour le Développement Agricole) which, in 1980/81, had outstanding a total sum of CFAF 50 thousand million. A large part of that sum was however related to commercial activities, and the total sum lended to farmers and groups of farmers amounted to about CFAF 15 thousand million. Of the total funds not less than 56 % was provided for coffee and cocoa plantations, against 22 % for cotton and 16 % for oil palm and coconuts.

Farmers can benefit from a subsidy for the rejuvenation of a coffee plantation, if their plantation is between 8 to 15 years old and has at least a density of 1000 plants per hectare. The amount payable is CFAF 60 000, of which a third is paid in kind (fertilizers, tools) and the remainder in cash, in two annual installments.

For various reasons farmers are stimulated to form cooperative societies (GVCs). These societies are assisted by the DCM (Direction de la Coopération et de la Mutualité). In 1980/81 there were about 1800 GVCs for coffee and cocoa marketing with a total membership of 84 000. These societies benefit of a so-called 'prime de groupage', if they collect the produce (coffee and cocoa) of their members. The total amount of coffee cherries and cocoa beans handled in 1980/81 was about 140 000 t, (15 % of national production) and the total payment received was CFAF 1.2 thousand million.

Apart from planting material, spraying equipment and chemicals and some fertilizers for rejuvenation of existing plantations, farmers do not require other inputs, and generally do not depend on the supply of inputs.

9.8 Summary and economic parameters

The area under coffee in Ivory Coast has expanded rapidly since the 1950s, and was in 1982 with a million hectares about as large as that in Colombia. But because of the low input cultivation methods the yields are very low, and although it is the third world-producer (competing for that with Indonesia), it produces much less coffee than Colombia and Brazil.

The country has hardly any high altitude zones and everywhere there is a warm tropical climate. Therefore the coffee grown is robusta, that is either grown in combination or competing with cocoa. Much research efforts have been devoted to the development of a disease resistant hybrid between arabica and robusta coffee, suitable for lowland areas. This resulted in the so-called arabusta coffee, for which a separate position on the world market is now being sought.

Whereas in the past both drying and hulling were undertaken at the village level, since the late 1970s 16 central hulling factories were established in the main

Economic parameters of the coffee sector in Ivory Coast (1982).

Resources used in coffee sector	
Land under coffee	1.0 million ha
as share of arable land	15 %
Labour involved	300 000 man-years
in production	260 000 man-years
in processing, marketing & supporting services	40 000 man-years
as share of national work force	7 %
Material inputs for coffee production	CFAF5 000 million*
Material inputs for coffee processing & marketing	CFAF18 000 million*
Annual import requirements for coffee industry	US$40 million*
Income from coffee sector	
Value of coffee production	
(volume × export unit value)	CFAF180 000 million
as share of agricultural production	23 %
as share of GDP	6 %
Value of coffee exports (f.o.b.)	US$510 million
as share of agricultural exports	28 %
as share of exports	22 %
Share of total production value, obtained by	
Producers	49 %
Processing industries, private traders & exporters	18 %
Government	33 %
Other parameters	
Robusta production	100 %
Yield (green coffee)	270 kg/ha
Production of labour involved (green coffee)	2.7 kg/man-day
Share of coffee produced for export	95 %
Ratio producer price (per kg green coffee) : daily wage rate (CFAF/man-day)	0.35

Sources: FAO (1971-1983), IMF (1983), World Bank (1984) and own estimates.

production zones. From these factories the coffee is transported to the premises of exporters in the ports of shipment (Abidjan mainly), where it is graded and bagged for export.

The parastatal 'Caisse de Stabilization' controls the marketing of coffee, fixes every year the producer price and marketing margins. In years of high prices on the world market it accumulates large funds, that are partly used for price stabilization purposes of the major export crops, and partly for the improvement of the rural infrastructure and rural institutions.

References

Asselman, G.M., 1980. Landendocumentatie 1979, no.2, Ivoorkust. KIT, Amsterdam.

Bonnefond, P. 1980. Statistiques Ivoiriennes-Secteur Agricole. ORSTOM, Paris.

Chauveau, J.P. & Richard, J. Une périphérie recentrée; à propos d'un système local d'économie de plantation en Côte d'Ivoire. Abidjan.
CSSPPA, 1984. Différentiel Café-Récolte. Abidjan.
Ediafric, 1980. L'agriculture ivoirienne. Paris.
FAO, 1971-1983. Production Yearbook. FAO, Rome.
FAO, 1980-1982. Trade Yearbook. FAO, Rome.
ICO, 1977-1983. Quarterly Statistical Bulletin on Coffee (several issues). ICO, London.
IMF, 1983. International financial statistics. IMF, Washington.
IRCC, 1982. Rapport d'activité. Paris.
Lee, E., 1980. Export-led rural development: The Ivory Coast. In: ILO/WEP study: Agrarian policies and rural poverty in Africa. Genève.
Les Atlas Jeune Afrique, 1983. Atlas de la Côte d'Ivoire. Editions J.A., Paris.
Marchés Tropicaux. April 1978. Le rôle du café dans l'économie de la Côte d'Ivoire. p. 1068-1074. Paris.
Ministère de l'Agriculture; Direction des statistiques rurales et des enquêtes agricole, 1981. Statistiques Agricoles 1981. Abidjan.
Priovolos, Th., 1981. Coffee and the Ivory Coast, an econometric study. Lexington books. Toronto.
Ruf, F., 1980a. Techniques culturales et productivité du travail en économie de plantation du Centre Ouest Ivoirien. GERDAT, Abidjan.
Ruf, F., 1980b. Note d'actualité: Le déterminisme des prix sur les systèmes de production en économie de plantation ivoirienne. GERDAT, Abidjan.
SERIC, 1977. Reconversion de l'usinage et du conditionnement du café. Abidjan.
Société Africaine d'edition. 1982. La Côte d'Ivoire en chiffres 1980-81. Paris.
Vaissière, P. de la, 1981. Structures et typologie des exploitations agricoles de la zone forestière de Côte d'Ivoire. INRA/CIRES, Abidjan.
Tuinder, B. van, et al., 1979. Ivory Coast. Challenge of success. World Bank, Washington.
World Bank, 1982. Staff appraisal report. Centre-West Agricultural Development Project. Washington.
World Bank, 1984. World development report 1984. Oxford University Press, Washington.

Currency equivalents and price indices in Ivory Coast (1960-1983).

	Exchange rate (CFAF per US$)	Consumer price index (1980 = 100)
1960	247	22
1965	247	25
1970	278	32
1975	214	46
1976	239	52
1977	246	66
1978	226	75
1979	213	87
1980	211	100
1981	272	109
1982	329	117
1983	381	124

Source: IMF (1983)

10 Coffee in Indonesia

10.1 Background

Indonesia is the most populous state in South-East Asia with 153 million inhabitants (1982). The total land area of 1.9 million square kilometres is spread over thousands of islands of which the great Sunda's: Sumatra, Sulawezi, Java and Kalimantan are the largest. Java is the most densely populated island with 450 inhabitants per square kilometre and a total population of 91 million. The population density on the other islands is much lower.

The GNP per person was US$580 in 1982. Agriculture contributes 26 % to the GDP and 58 % to total employment. The contribution of the other main sectors: industry and services to GDP is 42 % and 34 %, and to employment 12 % and 30 % (World Bank, 1984).

In the period 1980 to 1982 non-oil exports constituted about 30 % of total exports. To the exports of agricultural and forestry products, timber contributed about 40 %, rubber 24 %, coffee 12 %, palm oil 4 % and tea 3 % (FAO, 1980-1982). The agricultural area was in 1980 31.4 million hectares with 14.2 million hectares arable land, 5.3 million hectares under permanent crops and 12 million hectares permanent pasture (FAO, 1971-1983).

Table 10.1 gives the major crops in Indonesia. The Government's policy for the agriculture sector emphasizes food production, the creation of rural employment opportunities and the expansion of agricultural exports. Tree crops with its effects on food production and exports receive increased attention and extensive programmes are being implemented to improve production, processing and marketing facilities for smallholders as well as estates. The agricultural sector can be divided into two distinct subsectors: the smallholders and the estates. Within the smallholder sector the average farm size is less than 1 ha with rice as the main crop. Important cash crops for smallholders are spices, rubber, coconuts and coffee. The average size of estates is about 800 ha, and the dominant crops are: rubber, oil palm, sugar-cane, tea and coffee.

With an average annual production of 253 000 t, Indonesia contributed in the period 1979/81 about 5 % of the world production of coffee, and ranked fourth as world producer after Brazil, Colombia and Ivory Coast, and before Mexico and Ethiopia. It is the second world producer of robusta coffee.

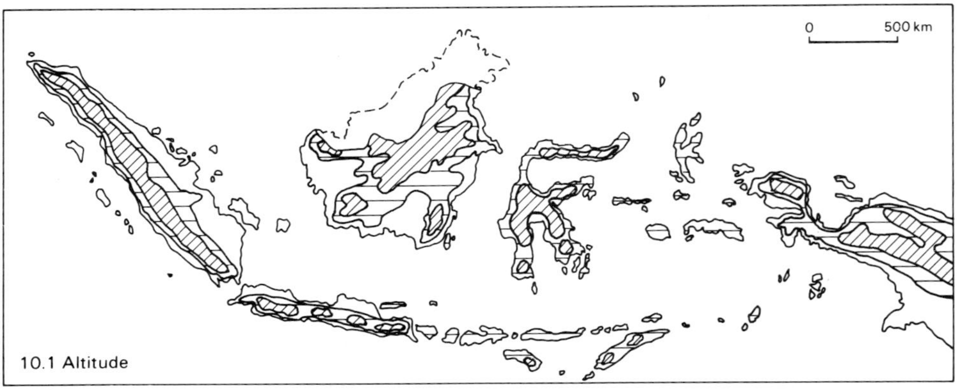

<100 m
100-1000 m
>1000 m

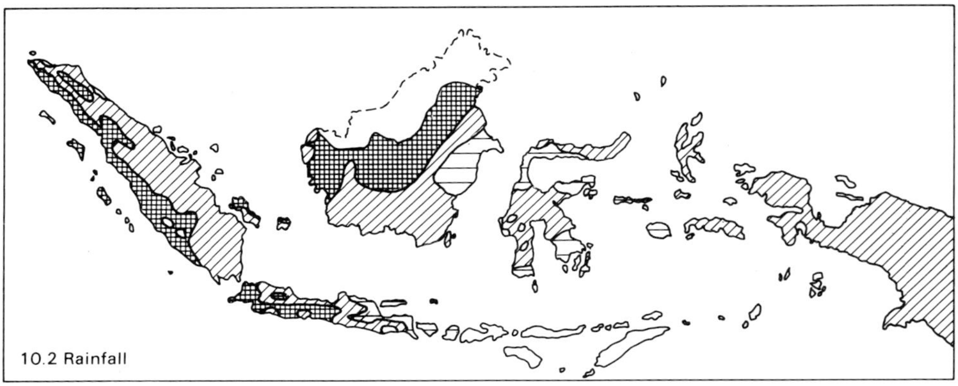

<1000 mm
1000-2000 mm
2000-3000 mm
>3000 mm

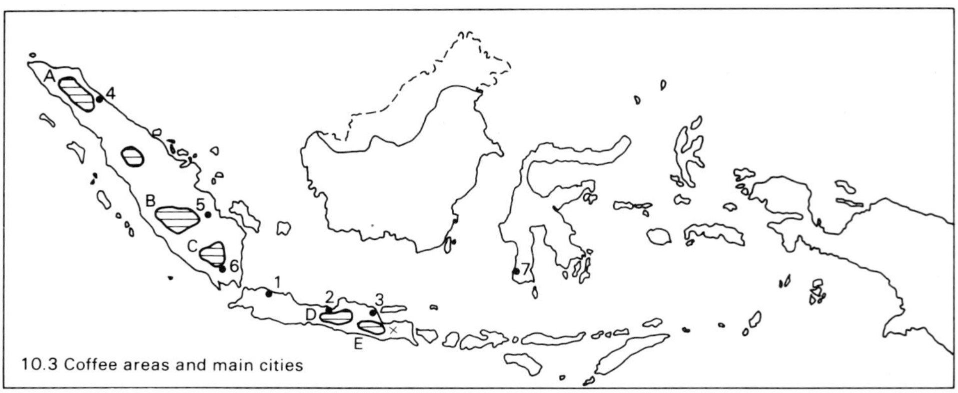

Coffee areas
A Aceh/N- Sumatera
B Bengkulu/S- Sumatera
C Lampung
D Central Java
E East Java

Main Cities
1 Jakarta
2 Semarang
3 Surabaya
4 Medan
5 Palembang
6 Tanjung Karang
7 Ujung Pandang

× Jember Research Station

Fig. 10.1-10.3. Altitude (10.1), average annual rainfall (10.2) and coffee areas and major cities (10.3) of Indonesia. After Lingen (1981).

Table 10.1. Average area harvested, yield and national production of major crops over 3-year periods.

Crops	Area (1000 ha)		Yield (kg/ha)		Production (1000 t)	
	1969/1971	1979/1981	1969/1971	1979/1981	1969/1971	1979/1981
Food crops						
rice/paddy	8158	8943	2346	3317	19136	29686
maize	2667	2700	965	1432	2575	3870
cassava	1425	1420	7512	9625	10695	13670
Oil crops						
oil palm	104	203	2567	4090	267	831
coconuts*	1700	2200	4314	4909	7333	10800
Beverages						
cocoa	9	18	184	490	2	9
coffee	380	409	455	617	173	253
tea	85	87	772	1069	65	93
Other crops						
rubber*	2200	2300	381	406	838	934
sugar-cane[1]	69	171	14014	9620	967	1645
tobacco	174	170	422	497	73	84

1. Production of sugar centrifugal and non-centrifugal.
Source: FAO (1971-1983)

10.2 Ecology

The country is composed of many islands, scattered over a wide area. Hence climatic conditions vary greatly. However all islands are within 10° of the equator, and temperatures are generally high (24-30 °C), without major fluctuations. The tropical rain forest climate is predominant. But on all major islands mountain ranges occur with tops over 3000 m and the mean temperatures decrease with altitude: about 0.6 °C for each 100 m. Therefore optimum altitudes are 800-1500 m (with temperatures between 19-21 °C) for arabica coffee and 400-800 m (21-24 °C) for robusta coffee (Figure 10.1). In most regions average annual rainfall is well over 2 000 mm and the major rainy period is usually during the west-monsoon (around January, Table 10.2). During the east-monsoon (July) the southeastern part of the archipelago experiences a dry spell of at least three months, with less than 60 mm rain per month. The average annual rainfall in those areas is less than 2000 mm (East Java, Bali) or even less than 1000 mm (Sumba, Timor) (Figure 10.2). The west coast of Sumatra, the south-west coast of Java and the northern part of Kalimantan, on the other hand, receive more than 3000 mm rain per year, without a distinct dry period, and are generally too wet for coffee.

With the exception of Kalimantan all islands do have fertile volcanic soils. On

Table 10.2. Distribution of rainfall in coffee areas.

Area	Station[1]	Annual rainfall (mm)	Monthly rainfall (mm)					
			J	F	M	A	M	J
Aceh	Takengon	1731	167	120	182	165	120	58
Lampung	Gsaat Nator	1918	227	313	234	128	247	72
E. Java	Gunung Malang	2241	303	253	261	167	170	150
			J	A	S	O	N	D
	Takengon		53	83	143	196	220	224
	Gsaat Nator		86	69	123	99	118	202
	Gunung Malang		156	142	136	113	153	237

1. Average annual temperatures respectively 19.9, 26.4 and 23.3 °C. Altitude 1205; 85 and 550 m respectively.
Sources: de Rijk (1981), Lambrecht (1982)

Sumatra the best soils are found in the north east and in the south in Lampung, both of which are important coffee areas. Taking into account the ecological conditions and the available land in the country, cultivation of coffee could certainly expand, particularly on Sumatra, Sulawezi and West Irian.

10.3 The coffee sector

10.3.1 History

Coffee (arabica) was first introduced in Indonesia around 1700, when the Dutch brought it from Yemen. In the 18th and in the beginning of the 19th Century it was mainly grown by the native population of West Java, and was exported through the Netherlands East Indies Company (OIC) It had a good reputation on the world market as 'Java coffee'.

In the period 1811 to 1816 a plantation system was introduced by Raffles, in which land concessions were given to private enterprises to develop estates employing indigenous labour, but in 1830 the Dutch Government decided in favour of the so-called 'Cultuur stelsel' with forced delivery of export produce to the OIC. The basis of this system was that the Javanese were to cultivate one fifth of the village's arable land with export crops designated by the Government, or work 66 days a year in Government owned estates or enterprises. In return, they received a small payment and a refund of land taxes (Sie Kwat Soen, 1968). Coffee, sugar, indigo and tobacco were then the major export crops. In 1870 it was made possible again for private enterpreneurs to acquire land for the large scale cultivation of export crops and the number and total area of estates increased rapidly during the period 1875 to 1937. By 1885 Indonesia was producing over 60 000 t of coffee a year and was the second largest producer in the world after Brazil. Then leaf-rust attacked the plantations reducing production by 1900 to only 25 % of former levels. Experiments with liberica coffee were not successful, but through the introduction of robusta, Indonesia became again a major world producer during the 1920s. World War II had a negative effect on production, and it was not until 1952 that the pre-war figure of 110 000 tonnes a year was exceeded. Since then, the coffee production increased steadily at an annual rate of at least 5 % to reach 160 000 t in 1969 and 320 000 t in 1982. This increase mainly results from increased plantings by smallholders.

10.3.2 Recent developments

The Tables 10.3, 10.4 and 10.5 show the development of the coffee sector in Indonesia since 1969, as far as area planted, production and the regional distribution of them are concerned. On the basis of this information, and other data, the following observations can be made, to characterize the present situation of the coffee sector in Indonesia:

– Coffee in Indonesia is predominantly a smallholder crop, and the area planted increased during the 1970s only for the smallholder sector. The area of estates remained the same.

– Through the 1976-coffee boom, following frost in Brazil, a major expansion of area planted took place between 1976 and 1979, resulting in an accelerated growth of production in the period 1978-1981.

– The domestic consumption of coffee remained stagnant over this period, with

Table 10.3. Area planted to coffee (1000 ha).

Crop year	Government estates	Private estates	Small-holders	Total
1969/70	20.2	22.1	337.3	379.6
1970/71	20.4	23.4	351.1	394.9
1971/72	20.6	20.7	366.3	407.7
1972/73	19.3	19.0	356.9	395.1
1973/74	21.0	19.7	340.5	380.2
1974/75	20.4	19.4	346.7	386.6
1975/76	20.1	18.4	361.3	399.9
1976/77	19.9	18.2	402.0	440.0
1977/78	20.3	23.3	454.3	497.8
1978/79	21.8	21.7	477.1	520.2
1979/80	20.8	25.8	582.4	629.0
1980/81	20.9	23.3	604.2	648.3
1981/82	20.9	23.4	627.1	671.3

Source: Central Bureau of Statistics (1978)

Table 10.4. Coffee production and average yield for areas harvested.

Crop year	Production (1000 t)			Yield (kg/ha)
	arabica	robusta	total	
1970/71	10	141	151	545
1971/72	10	108	118	393
1972/73	10	137	147	500
1973/74	11	149	160	562
1974/75	12	151	163	553
1975/76	14	174	188	586
1976/77	14	189	203	564
1977/78	15	224	239	613
1978/79	15	272	287	638
1979/80	14	274	288	600
1980/81	13	309	322	601
1981/82	14	333	347	630
1982/83	14	301	315	575

Source: USDA (1983)

Table 10.5. Proportional distribution (%) of area planted and coffee production by region (1970 and 1978).

Islands and provinces	Area (ha)		Production (1000 t)	
	1970	1978	1970	1978
Sumatra	55.4	61.9	69.1	76.4
Aceh	8.3	6.3	7.1	6.5
Benkulu	6.2	6.5	7.3	4.9
S. Selatan (South)	17.2	16.1	25.1	20.2
Lampung	13.8	22.9	16.5	36.4
Java	27.3	18.7	21.9	14.8
J. Timur (East)	14.6	12.6	12.2	10.3
J. Tengah (Central)	10.3	3.8	4.1	3.2
Sulawezi	6.6	7.0	4.1	3.9
Bali	7.0	6.2	2.5	1.8
Other islands	3.7	6.2	2.4	3.1

Source: Ministry of Agriculture (1980)

consumption per person becoming less than 0.5 kg.

– An accummulation of stocks, combined with a large proportion of export sales at low prices to non-ICO-members, took place in the period 1968 to 1973 and from 1981 onwards.

– The provinces of Lampung and Sumatra Selatan are the main production areas, and the expansion of the coffee sector took mainly place in Sumatra. The coffee production of Java, where most estates are situated, remained stagnant.

– Average yields increased hardly during the 1970s (nor during the 1960s). In 1981/82 average yields in Sumatra were about 600 kg/ha, against 400 kg/ha on Java, and lower yields on the other islands.

– Since the general attack of leaf-rust on arabica coffee around 1900, Indonesia is mainly a producer of robusta coffee and is classified as such on the world market. However still 5 % of the total coffee production is arabica, which is mainly grown in the Aceh province of Sumatra.

10.3.3 Organization

Whereas coffee is predominantly produced by smallholders, the marketing is entirely in the hands of private traders and exporters with the Government exercising a supervisory role only. The Directorate General of Estate Crops of the Department of Agriculture is responsible for planning, research and extension services within the coffee sector, and the Directorate General for Foreign Trade, within the Department of Trade and Cooperatives is responsible for quality control, testing and supervision of coffee exports. Through its estates the Government is to some extent (about 3 % of national supply) also directly involved in the production and marketing of coffee.

Because of complaints about insufficient quality, wet coffee and deliberate mixing, supposedly by middlemen, local government in a few regions once estab-

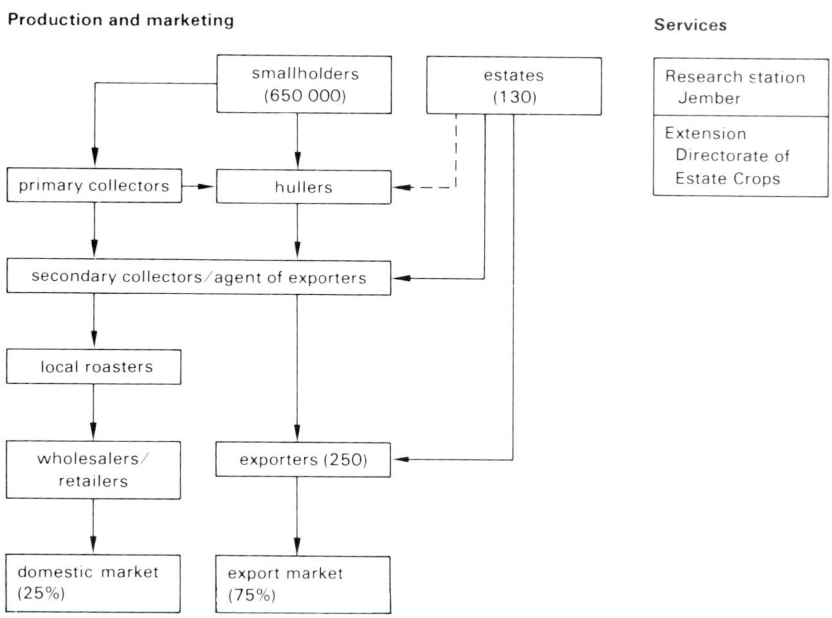

Fig. 10.4. Participants in the coffee sector and main marketing channels in Indonesia in 1981. Solid line, main channels; broken line, secondary channels.

lished coffee control measures, which were quite effective. But these were later abolished. The lack of adequate quality control measures remained a major problem, and affected the reputation of Indonesian coffee abroad. Therefore the Association of Indonesian Coffee Exporters was established in 1979. It aims at closer cooperation between trading parties and at improving quality and trade practices.

Since the development of the estate sector, from 1870 onwards, much attention has been given to agricultural research. Private experiment stations were established by the respective estate corporations, one of which, in D'jember (East Java), is still engaged in coffee research. The scientific results, in the past published in the annual publication 'Archief voor de koffie cultuur', enjoyed special fame, throughout the world. In Figure 10.4, the main participants in the coffee sector and marketing channels of coffee are shown.

10.4 Production

10.4.1 Production units and farming systems

The average size of smallholdings growing coffee is about one ha, ranging from 1.94 ha in Bali down to only 0.30 ha in Central Java. In the major coffee growing areas about 70 % are rather compact plantations with more than 1000 trees per hectare (average 1530 trees in 1974). The remainder consist of low density and small plantations with presumably about 500 trees per hectare. Coffee is also planted in forest areas.

In Java coffee is largely grown in mixed gardens, with both annual and perennial crops (pekarangan) or in forest gardens (mainly perennial crops and forest species). The forest gardens with a high planting density provide a good protection against soil erosion. Coffee is in Java often found in association with coconuts, bananas, pepper and tubers. In southern Sumatra (Lampung) smallholder coffee gardens, many of which are rather old (25 years and older), are frequently intercropped with cloves. Coffee is also found in combination with pepper plants climbing on supporting trees such as kapok and more recently *Albizzia* spp. Cloves and pepper are in that area also the major competitive crops to coffee, followed by tobacco and food crops such as rice and sweet potatoes.

In 1978 there were 129 coffee estates in Indonesia, of which 94 in East Java and 27 in Central Java. The estates range in size from about 30 ha to 1500 ha, with an average of 243 ha. The density is usually about 1500 trees per hectare. In 1938 there were still about 320 coffee estates and the total area planted and the production were twice as high as now. Estate coffee plantations were not always pure-stand plantations, for instance in Aceh coffee was interplanted in between such permanent crops as rubber, coconuts and oil palm, which have a long unproductive period. The coffee was often cut when the main crop reached maturity.

10.4.2 Production stages

New smallholders' coffee plantations were in recent times mainly established in Sumatra, usually on newly or recently cleared land. After clearing the land, usually upland rice is grown for two years, before coffee is planted. The unshaded coffee, is initially interplanted with food crops. Clearing and landpreparation takes on average about 120 man-days per hectare, planting 30 man-days per hectare and annual maintenance before the first major harvesting (after 3 years) about 50 man-days per hectare. The total labour requirements for the establishment of coffee therefore amount to 300 man-days per hectare. Replanting on the other hand requires about 210 man-days per hectare, and stumping, for the rejuvenation of a plantation, about 130 man-days per hectare.

Standards of cultivation by smallholders are generally higher in East Java than elsewhere, since methods are based on those used on neighbouring estates. The trees are regularly pruned, and normally provided with shade. Weeding is carried out regularly and a relatively large proportion of farmers (30 %) is using fertilizers, partly organic. In contrast farmers in Aceh (N. Sumatra) hardly use any fertilizers, weed irregularly and only about half of them prune. But in Lampung (S. Sumatra) through the activities of the PIK-programme, some farmers apply considerable amounts of fertilizers (300 kg urea, 125 kg TSP and 250 kg KCl per hectare) and insecticides (4.5 l Thiodan). In that area the agobiado method of pruning is recommended for the robusta coffee. On the estates coffee is normally grown under shade from *Erythrina lithosperma* or *Leucaena glauca*, the trees are regularly pruned, (single stem method) and chemical fertilizers are applied at a rate of 1 tonne of NPK per hectare.

Whereas smallholders in southern Sumatra usually apply strip cropping, followed by dry processing, the arabica growers in Aceh and the estates and smallholders in Java often apply selective picking and wet processing. The difference between the methods applied in South Sumatra and Java is probably related to the availability of labour and the cost of hired labour, which are higher in Sumatra because of the relatively sparse population. Whereas in Lampung (S. Sumatra) about 20 kg green coffee equivalent is harvested per man-day, arabica growers in Aceh were found to pick on average 39 kg cherries per day, or about 8-9 kg green coffee equivalent. This amount depends on the period of harvesting and on the extent that selective picking is applied. It can be assumed that under the conditions on Java, where selective picking is applied and where pruning facilitates picking 8 kg green coffee equivalent will be harvested per man-day. The main coffee harvesting season for most production areas is between June and December.

Yields on a smallholding in Sumatra in newly cleared forest may be initially about 900 kg/ha, but decline rapidly. A normal yield for smallholders is about 600 kg/ha and on poorly managed farms as low as 300 kg/ha. In Aceh (N. Sumatra) average yields of arabica were found to be higher than those of robusta (620 against 480 kg/ha). On well managed estates, yields are about 1000 kg/ha but average yields on estates, according to official statistics, do not differ much from average yields on smallholdings (about 530 kg/ha). Possible explanations for the relatively unfavourable input/output ratio for estates might be the ecological conditions (low rainfall) in some areas (East Java and Bali), excessive shade and low standards of management on certain estates.

The distance between the coffee fields and the farmers houses is often at least 0.5 km (and more on Sumatra) over which distance coffee is usually transported by foot, sometimes by bicycle. At the homestead the cherries are usually dried on mats or on the floor, and sometimes the pulp is first crushed to speed up the drying process. They are left in the sun for 3 to 7 days before being milled by hand with a pestle and mortar. The resulting hulled coffee, ('asalan') is usually sold without any sorting, and often at too high a moisture content (18 %).

In Aceh the cherries are kept 2-3 days (too long) in a tun or in bags before being pulped in hand pulpers with an average capacity of 150 kg/hour. Subsequently the parchment is stored under water for 12-20 hours, washed 1 or 2 times and dried (3-7 days) on mats. The coffee is afterwards stored at the house for an average period of 3 months. The farmers sell the dry or wet parchment coffee either at their home to neighbourhood collectors or village agents, or they have it hulled first and sell the green coffee subsequently to district traders, large traders or the huller in the village or main town. Some of the larger traders are in fact representatives of exporters. The parchment is brought to the huller by pickup or small bus (bemo), at a rate of about Rp 250 per bag or Rp 3 per kilogram (for the average distance of about 10 km). In Lampung transport costs ranged from Rp 100-500 per 100 kg of dried cherries. Farmers often obtain credit from the buyer, on average for about 5 months (interest 5 % per month).

10.4.3 Production costs and returns

As shown before, a broad distinction can be made between production units and production systems of smallholders in such different areas as North Sumatra (Aceh), South Sumatra (e.g. Lampung) and Java, and estates, most of which are situated in East Java.

Table 10.6 gives the annual costs of production for the production systems in those 4 regions. The costs made during the establishment period are added in the form of an annuity at 10 % interest rate over an estimated productive period of 20 years. The table shows that because of lower wage rates, the more intensive cultivation systems in Lampung and on the estates, incur approximately the same total costs per hectare as the production systems in Aceh, where little use is made of material inputs. Because of the slightly higher average yields, the total costs per kilogram of green coffee are lowest in Lampung, and this has indeed been the region where the area under coffee has expanded the most rapidly.

During the boom period in 1978 and 1979, the prices received by farmers for green coffee (equivalent) were about Rp 950/kg or higher, but in the latter half of

Table 10.6. Production costs of green coffee (per hectare and per kilogram) for three smallholder areas and one estate area (1981).

	Smallholders			Estates
	Aceh	Lampung	C. Java	E. Java
Yield (kg/ha)	600	650	300	550
Plant density (plants/ha)	1400	1500	1000	1500
Wage rate (Rp/man-day)[1]	1625	1100	600	750
Annuities of establishment costs (10 %, 20 years) (Rp 1000)	70	56	39	54
Annual costs				
Labour inputs (man-days)				
weeding	20	49	40	40
pruning	2	20	10	14
fertilizing	3	8	10	13
spraying	–	2	–	4
harvesting	71	45	40	70
pulping	2	–	–	4
washing & drying	3	4	2	4
others	5	5	4	7
total	106[2]	133	106	156
Subtotal of costs (Rp 1000)	172.2	146.3	63.6	117.0
Material inputs (Rp 1000)				
fertilizers	5.0	30.0	10.0	50.0
chemicals	3.0	6.0	3.0	15.0
bags & mats	1.0	1.0	1.0	–
other	2.0	2.0	2.0	6.0
transport fees	1.7	2.0	1.0	–
hulling fees	3.6	9.0	–	–
Subtotal	16.3	50.0	17.0	71.0
Overhead costs (Rp 1000)				
depreciation & interest of equipment	4.0	6.0	4.0	15.0
interest on loans	2.5	4.0	3.0	4.0
management	–	–	–	15.0
others	2.0	3.0	3.0	6.0
Subtotal	8.5	13.0	10.0	40.0
Total costs per hectare (Rp 1000)	267.0	265.3	129.6	282.0
Costs per kilogram (Rp/kg)	445	408	432	513
(US$/kg)	0.70	0.65	0.68	0.81

1. Wagerates in Lampung ranged from 700-1500 Rp per man-day, depending on activity.
2. About 30 % hired labour.

Sources: de Rijk (1981), Lambrecht (1982), Hunink (1984)

1981 prices had fallen below Rp 500/kg. At those 1981-coffee prices cloves, pepper and tobacco showed better net returns in Lampung province. For smallholders in East Java, cassava was estimated to give higher gross margins than coffee in that period. As long as prices of coffee on the world market continue to fluctuate and, without any price stabilization measures are passed on to farmers, and as long as they do not remain depressed over longer periods, it is likely that smallholders will continue to harvest at least part of their coffee, and will reinvest and increase inputs only in years of high prices.

10.5 Processing

The processing of coffee in Indonesia is largely decentralized. Depulping, drying and sometimes even hulling is carried out by the farmers themselves and therefore already described in Section 10.4.2. Sorting, grading and packing are usually undertaken by the exporters (Section 10.6.2).

Although a large proportion of farmers carry out the hulling themselves, specialized hullers usually perform the job faster and better (less damage). A survey among the 17 hullers in Central Aceh gave the following picture of the situation in that district; the average turnover per year was 125 tonnes of green coffee, against an average hulling capacity of 587 kg/hour. The engines used had an average power of 40 hp (about 30 000 W). During the main and second harvesting period the equipment was on average used for 9 and 2 hours per day, respectively.

About 60 % of the hullers owned a pick-up or truck and 68 % had a drying pad. Such a drying pad is important since farmers' coffee usually has a moisture content of about 18 %, whereas for hulling the maximum allowed is 16 %. Few hullers were operating stores, but village agents and farmers stored some 10 to 50 tonnes of parchment coffee for them. In Aceh the hullers generally receive 1 % of the product (green coffee) as payment for the processing, against 2.5-4 % in Lampung. There are no farmers' cooperatives engaged in coffee processing.

10.6 Marketing

10.6.1 Marketing channels

A national coffee marketing board or similar organization does not exist in Indonesia and neither are there small farmer marketing cooperatives. Therefore, and because of the large distances involved, the marketing channels for coffee are complicated. A multitude of independent middlemen are handling coffee, which can be divided into neighbourhood and village collectors, district agents, large traders from main towns and traders that operate on behalf of the exporters in the port of shipment. In the estate sector the marketing channels are shorter, and large estates take care of the export themselves. An outline of the marketing channels has already been given in Figure 10.4.

10.6.2 Marketing functions

The main marketing functions are collection, transport, sorting, grading, packing, storage and shipping.

Whereas a smallholder might sell between 10-100 kg green coffee per week, during the harvesting season, a primary collector buys in the order of 1-5 tonnes coffee from different origin and of highly variable quality. Secondary collectors, agents for exporters and independent collectors who sell to exporters might have a turnover of 100-300 t a month, which they transport over distances of 200-300 km.

Sorting and grading is carried out by the exporter. In 1980 there were about 250 licensed exporters of coffee. Most but not all of these have the necessary equipment for drying, cleaning, polishing and grading coffee. Whereas large exporters by now use electronic sorting equipment (with a capacity of 700-1500 kg/hour), numerous small exporters do not have any facilities at all. Many of the Aceh-licensed exporters do still use sorting by hand, for which they employ some 50 people (women) who can each sort about 120 kg green coffee per day. Including drying the total costs of hand sorting were in 1979 about Rp 15 per kilogram. Since the 'asalan' coffee received by exporters and their agents is usually too wet and below the minimum standard, losses are incurred in bringing this coffee up to the standard required. Packing is done in double bags of jute, at a cost of about Rp 660 per 60-kg bag (1979).

For a long time a system was applied, in which green coffee beans were first classified as OIB or GB ('Oost-Indische' or 'Gewone Bereiding': equivalent to dry processing) and WIB ('West-Indische Bereiding', or wet processing), respectively. Dry-processed coffee was then divided into grades according to the degree of sorting and polishing, and to a certain maximum triage (proportion of abnormal beans). Two separate grades for arabica were included. For wet-processed beans only two grades were distinguished according to the triage. Besides first grade robusta coffee beans (WIB 1) were classified into 3 sizes, by using screens with round holes of 5.5 and 6.4 mm diameter, respectively. The maximum moisture content allowed was always 14.5 %.

Under this system a substantial portion of Indonesian coffee did obtain export prices which were 5-20 % lower than those commended by other exporting countries. Therefore a new quality grading system was introduced, which became effective as from October 1983 and which has the following major characteristics:

- There are six grades, determined on the basis of a maximum number of defects (11 for first grade, 12-25 for second grade, etc.).
- For dry-processed coffee beans the maximum moisture content should be 13 % and dirt content 0.5 %; normal and large beans (screen 5.6 mm) are distinguished.
- For wet-processed coffee beans the maximum moisture content should be 12 % and dirt content 5 %; small, medium (6.5 mm screen) and large (7.5 mm screen)

beans are distinguished).

Not only farmers but also middlemen do store coffee for several months, usually as parchment coffee, and this evens out the supply to exporters, which, at least in Sumatra appears to be regular, given the definite harvest periods in major production areas. Exporters store also coffee in production areas but their major stores are situated in the ports of shipment, such as Teluk Betung, Surabaya and Palembang. The total capacity of warehouses in Indonesia is estimated at 330 000 t, or equivalent to about one year production, but the turnover is usually fast. Except for the period 1970-1972 and in recent years, stocks hardly exceeded the estimated required working stocks of 400 000 bags (24 000 t).

10.6.3 Pricing and taxation

The Indonesian Government intervenes little in the coffee trade, which is only to a limited extent affected by taxes and levies. In 1979, with international coffee prices at a peak, a new export tax was introduced to bring prices more in line with those of other export commodities and to keep down the local consumer prices of coffee. But there exists no price stabilization system or fund. In 1979 the following taxes had to be paid by exporters:

- turnover tax of exporter (2 % of check price)
- sales tax of farmer (Rp 10 per kilogram)
- turnover tax of farmer (Rp 8.15 per kilogram)
- regular export tax (5 % of check price)
- new export tax (18.28 % and 16.92 % respectively of the check price per kilogram of arabica and robusta coffee). The check prices stood then at US$3.58 and US$2.83, respectively.

Part of these tax revenues were earmarked for agricultural development in the respective regions, to be distributed by the central Government. The check prices were raised in June 1980 to US$4.15 per kilogram arabica, and US$3.20 per kilogram robusta, and since export prices remained below those figures, no export taxes were paid. In 1982 the export taxes were abolished. Import taxes for entry into the countries of the European Common Market are 5 % of the f.o.b. value at Indonesian port of shipment.

10.6.4 Marketing costs and margins

Table 10.7 gives a breakdown of export prices of coffee in different periods, and concerning the export of arabica coffee to Europe and robusta coffee to the United States, respectively. Table 10.7 shows clearly how fluctuating world market prices influence the domestic price level, even although the export taxes imposed during the period of high prices on the world market had a tempering effect on producer prices. Marketing costs and margins during this period (1980/81) were about Rp 500-600/kg or around US$0.90/kg.

Table 10.7. Prices, marketing costs and margins in Rp/kg[1] (green coffee equivalents), shipped from Medan to Europe or the United States, for three different periods.

	Febr. 1980, arabica, Europe		July 1981, robusta EK 1, United States		Dec. 1981, robusta EK 1, United States	
	(Rp/kg)	(%)	(Rp/kg)	(%)	(Rp/kg)	(%)
Ex-dock price	2446		1072		1394	
import tax	111		–		–	
C.i.f. price	2335		1072		1394	
freight, insurance, etc.	114		119		119	
F.o.b. price	2221	100	953	100	1275	100
Costs of export						
margin exporters, and losses during shipment	166	7	132	14	250	20
port and handling charges, interest, etc.	49	2	85	9	85	7
turnover & sales taxes	77	4	70	7	70	5
export taxes	504	23	–	–	–	–
Total	796	36	287	30	405	32
Cost of marketing						
sorting, drying, bags	25	1	30	3	30	2
losses (moisture, dirt, triage)	150	7	70	7	80	6
processing costs (hulling)	20	1	10	1	10	1
transport	20	1	20	2	20	2
costs and margins collectors	70	3	70	7	105	8
Total	285	13	200	20	245	19
Price to growers	1140	51	466	49	625	49
Costs of marketing and export (excluding export taxes)	Rp577/kg US$0.92/kg		Rp487/kg US$0.77/kg		Rp650/kg US$1.03/kg	

1. Exchange rate in 1980: Rp 627 per US$ and in 1981: Rp 632 per US$.
Source: de Rijk (1981)

10.6.5 Export and domestic consumption

Little information is available on the domestic consumption of coffee. Taking into account production and export statistics, the average per person consumption of coffee in Indonesia is around of 0.5 kg. The domestic coffee is usually made up of non-exportable produce, products from triage such as black beans etc., and often blended with roasted areca nuts and maize, the price of shelled maize being only about 10 % of local prices of pure coffee.

Local supply and local coffee prices fluctuate with prices on the world market and during periods of high prices on the world market the local brew usually contains less coffee and more substitute materials. The wholesale price for coffee (robusta) in the Jakarta market was about Rp 320/kg in the period 1973/75, before the increase of prices on the world market. It increased up to Rp 800/kg in 1976, and was as high as Rp 1000/kg in January 1981, after which it dropped to about Rp 700/kg (early 1982).

Indonesian coffee producers and exporters have reacted fast on the rising prices on the world market in the period 1976-1979, and in 1978/79 Indonesia exported twice as much coffee as in 1974/75 (Table 10.8). Since then, exports remained at about the same level, whereas stocks started to accumulate since 1982.

There are about 250 registered and licensed exporters of coffee in Indonesia, but only 25 of those handle large quantities. The exporter is the prime mover in the marketing system as he normally provides the finance to enable agents and collectors to buy coffee. Being located normally some 200 km from the coffee growing areas and dealing with agents, the exporters, at least in Sumatra, have little contact with the growers and have no first hand knowledge of the factors affecting the quality of the 'asalan' purchased. A survey among coffee exporters in North Sumatra (Aceh) showed that the average time required for collection (7 to 8 days), sorting and drying (6 days), arrangements for shipment (7 days) is only about 21 days, which is also the average period coffee is stored by exporters. The average quantity for one shipment was about 10 t. The exporters employed on average about 10 permanent labourers, but half of their turnover consisted of other produce, e.g. rubber, wood and copra.

To become an exporter, a trader has to fulfill the following conditions:
– to possess a permit and a recognised export number, issued by the Provincial Trade Department
– to be listed as 'Registered Coffee Trader' by the Ministry of Trade (Director

Table 10.8. National production, domestic consumption and exports (in 1000 t) (1973-1982).

Crop year	Opening stock	Production	Consumption	Available for export	Export		
					total	ICO members	non-members
1973/74	27	160	59	128	95	92	3
1974/75	33	164	61	136	106	104	2
1975/76	30	187	62	155	135	132	3
1976/77	20	203	60	163	140	140	0
1977/78	23	239	62	200	176	175	1
1978/79	26	288	65	249	225	224	1
1980/81	24	322	74	272	232	230	2
1981/82	40	347	70	317	209	150	59
1982/83	108	315	69	354	231	.	.

Sources: USDA (1983), ICO (1977-1983)

Table 10.9. Proportional distribution (%) of coffee exports to major importing countries (1969-1981).

	Four years periods		
	1969/70-1972/73	1973/74-1976/77	1977/78-1980/81
ICO members			
United States	50	43	31
Netherlands	14	11	13
Italy	3	4	13
Japan	3	8	8
Singapore	1	4	7
France	4	3	7
West Germany	1	2	4
others	12	25	16
Non-members	12	2	1

Source: ICO (1977-1983)

General of Export Trade)
– to own a storage and sorting plant, and to have sufficient capital, on a Government or Foreign Exchange Bankaccount
– to have experience in the coffee trade and (e) to obtain for each specific period a basic quota from the Ministry of Trade.

Major ports of shipment for coffee are Surabaya, Semarang and Jakarta for Java; Palembang, Benakulu and Medan for Sumatra, Bulelang for Bali, and Ujung Pandang for Sulawezi. Only three of these were in 1982 container ports: Surabaya, Jakarta and Medan, whereas shipments of coffee to the United States and Europe were containerized in 1979 and 1982, respectively. Costs of sea transport (from Medan) were in 1980 as follows: to Japan US$0.05/kg; Europe US$0.12/kg; to the United States East-coast US$0.18/kg and West-coast US$0.23/kg. For tea, rubber and oil palm regional markets (Pakistan, Australia and Japan) are relatively more important than for coffee, which apart from sales to Japan and Singpore, is mainly exported to the United States and Europe. But during periods of low prices on the world market (1968-1970, 1981/82) exports to countries outside ICO were considerable (e.g. North Africa, China). In recent years exports were more diversified (Table 10.9).

10.7 Supporting services

As far as the smallholder sector in Sumatra is concerned, linkages with the other sectors (industry and services) are generally not important: farmers use relatively little material inputs (fertilizers, etc.), they do not obtain much credit from institutional sources (from buyers instead), and they carry out at least part of the processing themselves, with simple implements. Cash is required mainly for the purchase of pulpers, new coffee plants, sacks and other materials.

In 1979 the Directorate of Estate Crops established some production packages for the promotion of export crops among smallholders. These packages were meant to increase farmers' income through higher yields, to contribute to an expansion of the total area under industrial crops and to boast the foreign currency income of the country. They were established for rubber, local and hybrid coconuts, cocoa, tea, pepper and coffee, and were made attractive for the farmers through provision of a certificate of landownership. The packages were handled through the offices and branches of the Directorate of Estate Crops, in close cooperation with the extension services, the Ministry of Finance, BAPPENAS and the banks involved (usually the Bank Rakyak Indonesia).

For coffee, there were the packages of the 'Proyek Intensifikasi Kopi' (PIK) for the 'rejuvenation' and 'establishment' of coffee gardens for which credit could be obtained to a maximum of Rp 650 000 in 3 years (1979/80) at an interest of 6 % and grace periods of 3 and 4 years, respectively. Because of the low export prices since 1980, the expansion of the coffee area is no longer stimulated, and assistance is only provided for the intensification of coffee gardens.

The estate coffee production on Java can on the other hand be characterized by a relatively high level of inputs which the estates acquire themselves or through agents.

10.8 Summary and economic parameters

In the 18th Century, Indonesia was the first major coffee-exporting country and now still belongs to the four major exporters. Because of a disastrous attack of leaf-rust on its arabica coffee, the country switched around 1900 over to the production of robusta. Large areas are suitable for coffee and particularly on the the less densely populated islands there is considerable scope for the expansion of coffee production.

There still remain about 130 governmental and private estates, but the bulk of the coffee is produced by smallholders, on small plots, usually in forest gardens. Coffee processing (hulling mainly) is generally a small scale operation undertaken by traders and sometimes by farmers themselves. Sorting and grading is carried out by the exporters.

In Indonesia, coffee is not a major export commodity, and the Government does not intervene much in the marketing and trade of coffee. There are also hardly any cooperative organizations involved in the processing and marketing of coffee, and because of the very long distances there are many middlemen involved in the trade of coffee between producers and exporters.

Indonesia is basically a low cost producer but the marketing costs are high and because of an inadequate quality control, it generally fetched low prices on the world market. Besides its exports to countries of the European Community are charged with a 5 % import duty, whilst African robusta exporters are exempt from that duty.

Economic parameters of the coffee sector in Indonesia (1981).

Resources used in coffee sector	
Land under coffee	0.7 million ha
as share of arable land	3 %
Labour involved	360 000 man-years
in production	270 000 man-years
in processing and trade	90 000 man-years
as share of agricultural work force	0.5 %
Income from coffee sector	
Value of coffee production	
(volume × export unit value, f.o.b.)	Rp400 000 million
as share of agricultural production	3.1 %
as share of GDP	0.7 %
Value of coffee exports (f.o.b.)	US$420 million
as share of agricultural exports	8.6 %
as share of exports	1.9 %
Share of total production value, obtained by	
Producers	50 %
Traders and exporters	40 %
Government	10 %
Other parameters	
Robusta production	95 %
Arabica production	5 %
Yield (green coffee)	500 kg/ha
Production of labour involved (green coffee)	3.0 kg/man-day
Share of coffee produced for export	75 %
Ratio producer price (per kg green coffee):	0.45 (Sumatra)
daily wage rate (Rp/man-day)	0.85 (Java)

Sources: FAO (1971-1983), IMF (1983), World Bank (1984), and own estimates

References

Central Bureau of Statistics, 1978. Data statistik tanaman kopi. Jakarta.

FAO, 1971-1983. Production Yearbook. FAO, Rome.

FAO, 1980-1982. Trade Yearbook. FAO, Rome.

Huitema, W.K., 1935. Bevolkings koffiecultuur op Sumatra. Wageningen.

Hall, C.J.J. van & Korpel, C. van de (Ed.), 1946. De landbouw in de Indische archipel. Den Haag.

Hunink, R.B.M., 1984. The coffee development in Indonesia. Thesis for Agricultural University, Wageningen.

ICO, 1977-1983. Quarterly Statistical Bulletin on Coffee (several issues). ICO, London.

IMF, 1983. International financial statistics. IMF, Washington.

Lambrecht, Ph. C., 1982. Crop-economic considerations, Sekampung Watershed (Sumatra). FAO/INS/78/006, Bogor.

Lingen, J.P., 1981. Landendocumentatie 1981, no. 1. Indonesië. KIT, Amsterdam.

Ministry of Agriculture, 1980. Direktorat Jendral Perkebunan, Jakarta.
Rijk, P.A.L. de, 1981. The marketing and processing of coffee from development planning area III. IDAP, Aceh.
Sie Kwat Soen, 1968. Prospects for agricultural development in Indonesia. Pudoc, Wageningen.
World Bank, 1984. World development report 1984. Oxford University Press, Washington.
USDA, 1983. Foreign agriculture circular on coffee. USDA, Washington.

Currency equivalents and price indices in Indonesia (1970-1983).

	Exchange rate (Rp per US$)	Consumer price index (1980 = 100)
1970	365	20
1971	393	21
1972	415	22
1973	414	29
1974	415	40
1975	415	48
1976	415	58
1977	415	64
1978	442	69
1979	623	84
1980	627	100
1981	632	112
1982	661	123
1983	909	137

Source: IMF (1983)